PRINCIPLES OF CAD

Also by A. J. Medland
The Computer-based Design Process
CADCAM in Practice
(with Piers Burnett)

Also by Glen Mullineux
CAD: Computational Concepts and Methods

PRINCIPLES OF CAD

A Coursebook

A. J. Medland
Glen Mullineux

KOGAN
PAGE

First published 1988 by
Kogan Page Ltd
120 Pentonville Road
London N1 9JN
Copyright © 1988 A. J. Medland and G. Mullineux

British Library Cataloguing in Publication Data
Medland, A. J.
    Principles of CAD: a coursebook
    1. Engineering design —— Data processing
    I. Title   II. Mullineux, G.
    620'.00425'0285   TA174

ISBN 1-85091-534-2

Typeset by Saxon Printing Ltd., Derby

Printed and bound in Great Britain by
Billing and Sons Ltd., Worcester

CONTENTS

AUTHORS' PREFACE

Many books already exist on computer-aided design and manufacture most of which are dedicated to describing the complexities of mathematical modelling and its application to industrial problems. In the experience of the present authors, however, if the subject is to be understood within its true, industrial context it must be taught in relation to the design process. Thus, while this book discusses both modelling and industrial applications, it also tries to provide an insight into design methodology, system selection and usage, and the social relationships that exist within design and manufacturing facilities.

The teaching modules which make up the book are the distillation of material used by the authors both for undergraduate courses in CAD at Brunel University, and for seminars given to industrial users. The modules are not intended to be used in isolation, but rather to serve as an introductory survey which will enable students to grasp the broad outlines of the subject. Most aspects of the course presented here will need to be supported by further work and reading (see 'Further Reading').

In the authors' own courses much of the geometric and modelling work described in the text is supported by tutorial activities using the university department's commercial and research CAD/CAM systems. These include the Computervision CADDS4X and Personal Systems. While some of the examples given were prepared using these systems, care has been taken to present material in the broadest possible context in order that it may be easily adapted to other systems with similar capabilities.

The design and industrial applications material contained in the book is supplemented, during courses, by case studies and field work undertaken with local industrial companies. Details are given where possible. It is, however, strongly recommended that users of the book should endeavour to tailor such supplementary work to take full advantage of the opportunities and circumstances that exist in companies known to them.

Throughout the book both the industrial employees and the students are referred to as though exclusively male, this should be taken as only a reflection on our inability to devise a non-cumbersome, non-sexist alternative. It is not intended to suggest that it is our wish that these should be seen as masculine activities; for it is our experience that both sexes are represented within our student group and industrial contacts, and that they demonstrate equal abilities.

We wish to thank our colleagues for their tolerance during the preparation of this manuscript and to dedicate it to our respective parents.

Tony Medland
Glen Mullineux
January 1988

HOW TO USE THIS BOOK

The text is made up of a series of 'modules' which are grouped into eight sections. The intention has been, wherever possible, to make themodules entirely self-contained, so that they can be read in any order and the material they contain can be adapted to meet the needs of the widest possible range of readers. Figure 0.1 shows something of the interrelation of the sections.

The prime purpose of a CAD system is that it should enable a designer to use the computational and data-handling power of a computer to help him in his work. In order to appreciate how this can be achieved the student must understand something of the three elements which make up the term 'computer-aided design': that is, to say he must have a grasp of what the *design* process involves, of what a *computer* can do and of how its power can be enlisted to *aid* the designer in his task.

> **Design:** Section 1 provides an overview of the design process and Section 7 describes how the process fits into the organizational structure of an industrial company.

> **Computer:** Section 2 discusses aspects of computer hardware, from input devices to systems networking. Section 3 provides an introduction to the software that is used to construct geometric entities; Section 4 looks at how this software is used to present geometrical images; and Section 5 goes on to show how these methodologies are put together in a fully fledged CAD system.

Section 6 deals with the way in which a user interacts with a system.

Aid: Section 7 discusses how design aids are integrated into a company and the ways in which CAD impinges upon the organization of a company. Section 8 describes a number of application areas.

Naturally, in a book of this size it is not possible to go into any particular aspect of design in great detail, or to compare the various commercial CAD systems. The aim has been to keep the material as general as possible. The 'Further Reading' section at the end of the book provides lists of references for each section to assist the reader who wishes to find out more about any specific topic.

As already mentioned, the modules are more or less independent. Although they have been arranged in sections, it is not essential to read them in the order in which they are presented, although this would be a logical way through the text for someone wishing to cover all the material presented in the book. For readers wishing to concentrate upon particular topics some possible 'routes' are suggested below. Some routes are short and are designed to provide a brief overview which may, we hope, encourage the reader to delve into other modules in order to broaden his understanding.

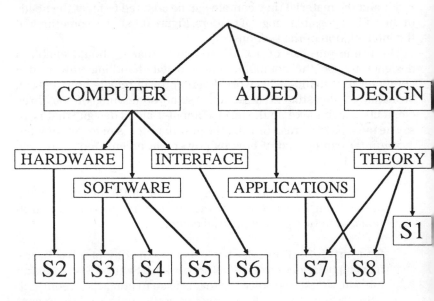

Figure 0.1

Route 1 – Quick introduction to what a CAD system looks like

Module 2.1
Module 2.2
Module 5.2
Module 6.1
Module 6.2

Route 2 – Quick review of how CAD systems work

Module 3.1
Module 3.2
Section 5

Route 3 – How CAD and geometric software operates

Section 3
Section 4
Section 5

Route 4 – Quick review of how CAD aids design

Module 1.2
Section 8

Route 5 – The impact of CAD on design

Section 1
Section 7
Section 8

SECTION 1
THE DESIGN PROCESS

It is perfectly possible to treat the mathematics and techniques of computer-aided design simply as one application of computer graphics. For there is no need to understand what a line represents or is to be used for in order to construct it upon a graphics display device. But to approach the subject in this manner is to adopt a blinkered point of view that concentrates upon the computer programming techniques involved and ignores the major benefits that such techniques can bring to industrial users.

The development of CAD has been driven by a need to solve growing problems in design. In order to ensure that the most appropriate techniques are applied, it is necessary to understand the nature of these underlying problems. Often the 'pure science' approach will lead to solutions that, while exact and elegant, are too complex to use in practice. The designer lives in an uncertain world of half-developed ideas, of constraints and compromises. On many occasions he is concerned less with achieving ideal solutions than with generating approximate values that can be used to bound his ideas.

Before embarking on a discussion of CAD it is therefore necessary to say something about the design process itself. This will provide a framework within which CAD techniques can be evaluated and compared. It will also allow the differing needs of various industries to be recognized and indicate why a whole range of CAD systems with different characteristics has been developed.

MODULE 1.1 THE RELATIONSHIP BETWEEN GEOMETRY AND FUNCTION

The work of the designer starts long before he approaches the computer-aided design terminal or the drawing board. It is perhaps only in an engineering drawing class, where the student may be presented with a complete set of instructions and details from which he is expected to construct a drawing, that a design activity is approached with a completely empty mind!

SPECIFICATION OF NEEDS

In more realistic situations the design activity commences with the specification of a set of needs. The item to be designed has a purpose, and the objective of the design process is to produce an object, or an assembly of objects, that satisfies those needs within the constraints inposed by commercial and other factors. The needs are usually defined in terms of the functions which the object is required to perform. If, for example, we are asked to design a bracket to support a floodlamp above a sports field (figure 1.1), then the need is for a means of maintaining the lamp in a fixed position and the bracket is required to provide those functional properties of strength and attachment which are necessary to support the weight of the lamp and fix it to a pole or tower.

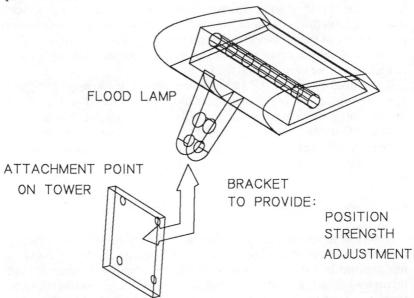

Figure 1.1 Functional needs of a mounting bracket.

DERIVING FUNCTIONS

The determination of the functions to be performed by objects, whether individually or as part of an assembly, can be thought of as the first stage in decomposing the design needs. But this process is itself to some extent dependent upon the designer having already fixed upon some provisional solution. We might, for example, be required to provide a means whereby the lamp may be redirected to illuminate a different area of the ground. In deciding whether of not the adjustment function should be attributed to the bracket, we will be influenced both by the details of the specification and by our own approach to the problem.

Should the requirement be that 'some adjustment is to be provided' so that the direction can be changed, then the inclusion of articulated linkages and/or a ball-end clamp that can be adjusted by a maintenance man with a spanner, may be all that is necessary (figure 1.2).

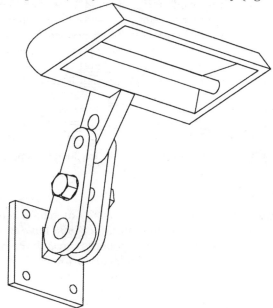

Figure 1.2   Ball-ended clamp and linkage for adjustment.

But if the requirement is that 'the lamp is to be able to follow play' a different solution will clearly be needed. Now the functional requirement of repositioning must be provided and attributed to one part of the overall design. The lamp may be turned by providing relative motion between lamp and bracket, bracket and supporting structure, structure and ground, or even by a combination of all three (figure 1.3). Whilst the outcome must be an adjustable lamp, the distribution of that function within the scheme is totally dependent upon the imposed constraints and the chosen design.

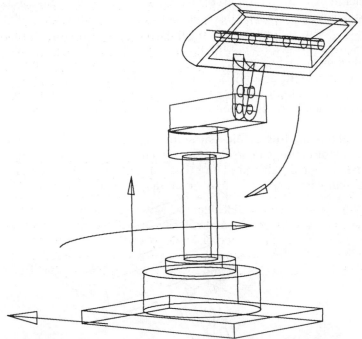

Figure 1.3   A combination of motions provided by separate devices.

GEOMETRIC FORM

Once the design has reached the stage at which the primary needs have been established and all functions have been individually attributed to assemblies or components through the development of a design scheme, the geometric interpretation can commence. As most design is conceived in visual terms it is unlikely that the designer will have arrived at this point without some fairly clear idea as to what the design will look like. But it is important to realize that up until this point imagery is used only as an aid to interpreting the functional requirements and not as a geometric description that must be defended. It is the ideas underlying the imagery that are important, not the form.

But once the designer starts to develop the geometry of the object he is constructing a single, specific embodiment of the functional form of the design. Should problems be encountered with one interpretation then he is, of course, still free (if the design commitment is low) to switch to another. And if additional constraints and functions are imposed then a new or modified scheme will need to be developed which may have far-reaching consequences.

It is the role of the design office staff to produce unambiguous engineering and geometric information from which objects may be produced and assembled. Each object detailed must have a purpose – or why else are we designing it? But it must also relate to other objects in order that its purpose may be satisfied. If our lamp is to be mounted upon a wall then the bracket must relate to both the lamp and the wall. The nature of these relationships depends upon the geometric shapes in the local regions as well as upon functional requirements such as strength.

Geometry-based Information

It is therefore clear that there is a close relationship between geometry and function. While function is the main driving force of design, it is geometry that is handled and manipulated throughout the manufacturing operations. Often, a designer or engineer will be provided only with geometric information and will be left to determine the function of an object from experience or by deduction. With some items the connection between function and form is obvious, as it is for a cam (figure 1.4) where geometry and function are the same thing. But if a

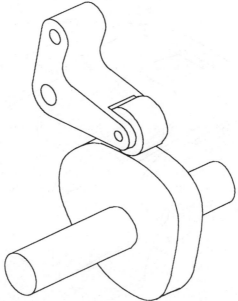

Figure 1.4 A cam and follower arrangement.

single component (figure 1.5) is isolated from the complex system of which it forms a part, then, while some features, such as mounting

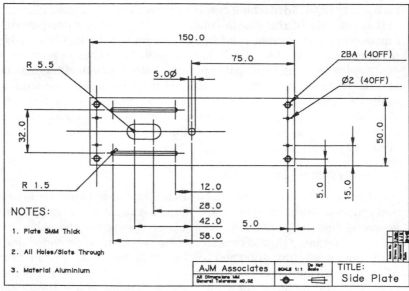

Figure 1.5 An engineering drawing of a single component.

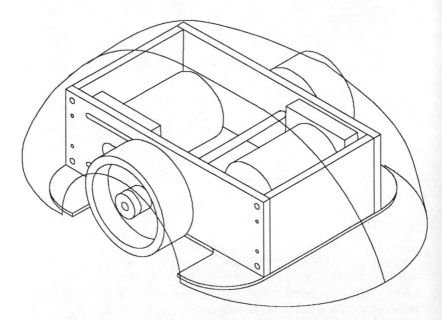

Figure 1.6 The component in Figure 1.5 shown in its working situation.

holes, may provide clues as to its purpose, its functions are likely to remain undetected until it is seen in its working situation (figure 1.6). Even then many of the more subtle or complex functions and their interrelationships maybe difficult to discern, even with an experienced eye.

It is, therefore, good practice to record the functions being fulfilled by all features of a design. The holes in the lamp bracket upright are there to allow it to be fixed to the wall with screws. It follows that the number of holes must be sufficient to fix it securely and that they must be of the appropriate size (perhaps they should also have been countersunk?). Each region of that component can thus be identified as contributing some attribute or function to the overall effectiveness of the bracket design (figure 1.7).

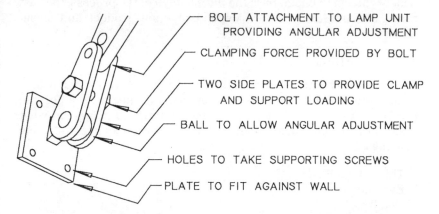

BOLT ATTACHMENT TO LAMP UNIT PROVIDING ANGULAR ADJUSTMENT

CLAMPING FORCE PROVIDED BY BOLT

TWO SIDE PLATES TO PROVIDE CLAMP AND SUPPORT LOADING

BALL TO ALLOW ANGULAR ADJUSTMENT

HOLES TO TAKE SUPPORTING SCREWS

PLATE TO FIT AGAINST WALL

Figure 1.7 The functional contributions provided by various elements of the bracket design.

MODULE 1.2 THE FORM OF THE DESIGN PROCESS

The process whereby objects are designed is seen to vary widely throughout the engineering industry. Many academics have taken this diversity as an indication that design is a multi-faceted activity which defies rational description. Others have set out to describe every elemental step in the design of a particular device or artefact and to arrange the steps in an enormous logic tree, thus reducing the process to one of structured decision-making and problem-solving.

Both these extreme views are unhelpful. The first approach sees design as general but unpredictable, whilst on the second takes it to be specific, and thus fully describable in all its detail. The truth lies somewhere in between. While the processes employed may not be fully understood they all contribute towards achieving the goal of producing the desired article. They are therefore logical but not necessarily efficient. They lead to success but not always by the easiest route. Some activities have even been known to deflect progress and make the target more difficult to achieve.

INTERDEPENDENCY OF DESIGN

The complexity of the design process arises from the fact that the activities involved are dependent upon both the product being designed and the social structures imposed upon the design staff. Thus, while an underlying framework, known as the *anatomy* of the design process, can be discerned, that framework must be drawn widely enough to encompass the *morphology* of the process, the vast range of variations that are dependent upon product and social factors.

ANATOMY OF DESIGN

All design must commence with the definition of a need which will, in turn, result in the generation of an agreed specification of the product's function. The activities which follow, no matter on what scale they are conducted, can be categorised under the following headings:

Think of an idea
Evaluate it
Detail it
Make it

to which we might sometimes like to add:

Control events
Test what we have made (to see if it is what we wanted)

Each of these six basic stages is to a greater or lesser extent, a part of every design process, although the terminology used to describe them may vary widely, with many words used interchangeably. There is also confusion regarding the definition of the word 'design' itself. Here it is used to describe the whole process from the original idea to the final test, but many would claim that the making and testing stages fall outside the scope of design (as in a 'design and make' project). Similarly, the definition of a 'designer' varies widely from company to company.

Here, the six stages, or blocks, which make up the design process have been termed *concept, analysis, scheming, manufacture, control* and *evaluation.* These are related as shown in figure 1.8 and we shall now consider each in turn.

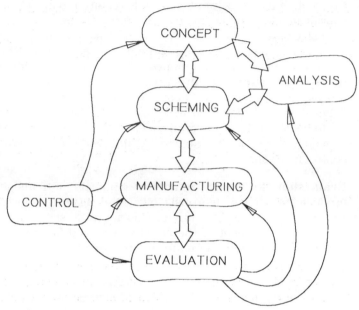

Figure 1.8   The design process.

Concept: This is the phase in which alternatives are considered and evaluated. Once a preferred solution or approach emerges the outline design is created and documented. It is this that is passed on to initiate the scheming stage.

Scheming: At this stage the design outline is turned into a single embodiment of the original idea, and all the engineering detail is added in order to describe the component parts and their assembly.

Analysis: This activity sits in the loop that connects the concept and scheming stages. Due to differing design requirements, the activities may either proceed from concept through analysis in order to provide the necessary details for scheming (clockwise motion in figure 1.8), or, alternatively, schemes may be developed directly from the concept before being analysed to check that they meet the original objectives (anticlockwise motion in figure 1.8).

Manufacture: This is the process whereby the geometric description generated at the scheming stage is translated into a physical product.

Control: Some measure of control is required to organize and coordinate all the preceding processes, but the need for control is most apparent during the scheming and manufacturing stages. As it is not always possible to complete one process in its entirety before another commences, there will be a considerable flow of information in all directions across the complete design process. For example, the design of elements that have been detailed and manufactured will impose constraints upon items yet to be designed. The overall success of the process will thus depend heavily upon the sensitivity and responsiveness of the control function.

Evaluation: No design activity is complete until the item produced has been rigorously evaluated against its original specification and design outline. This appraisal should include an honest assessment of the effectiveness and failures of the product, its expected useful life and its possible replacement.

MORPHOLOGY OF DESIGN

Whilst the design process will, in principle at least, involve all six stages, in any particular case the relevance or weight attached to each will be product-dependent. It is, therefore, sensible in this context to classify products, not by industry, discipline or level of technology, but in terms of the degree of constraint imposed upon the design. Some designs are easy because many solutions are apparent and the process is thus one of selection. Others are difficult because the requirements are many and conflicting and the problem thus reduces to one of optimization. These two extremes can be described as *under-constrained* and *over-constrained* design conditions.

OVER-CONSTRAINED DESIGN

The design of an advanced technology product, such as an aircraft, is over-constrained. The requirements of flight performance, payload, structural strength, economy and passenger comfort all impinge upon the design and frequently come into conflict with each other. In such cases the designers' problem is essentially one of finding an optimum solution and the design process will be heavily biased towards the analysis stage. As a result the emphasis of computer-aided design activity in the aircraft industry is firmly placed upon the use of large analysis programs for aerodynamic, thermal and structural calculations. The use of such programs allows alternatives to be evaluated and compromises reached by cycling around the concept-analysis-scheming loop.

So strong is the dependency of the over-constrained design upon this analysis iteration procedure, that on some occasions it will control, or even dictate, all other downstream operations. In order to achieve new levels of performance, meet new and more stringent regulations, or achieve new and more economical standards of manufacture, the traditional processes and manufacturing techniques may have to be abandoned. It has sometimes been found to be easier to build a new factory or production line rather than adapt the old one.

Over-constrained design thus centres upon the use of analysis to obtain a clear and accurate interpretation of the product in the scheming stage. The high quality product definition that results imposes rigorous demands upon the rest of the design activities.

UNDER-CONSTRAINED DESIGN

The design of an office chair must satisfy many objectives including strength, comfort, appearance, simplicity of manufacture and price. However, none of these requirements is so rigid as to over-constrain the process. Many acceptable solutions exist, as is apparent from the range of office furniture that is available. The choice of design is therefore governed by factors other than that of optimization by analysis. The technical requirements of supporting a person in a sitting position leave the designer with many options. Since the design process is under-constrained, the designer can select a solution on the basis of other considerations such as style, ease of manufacture and market economics. Depending on which consideration carries most weight, the emphasis in under-constrained design may fall upon any one of three stages – manufacture, scheming or concept.

Under-constrained, manufacturing-centred: If a company is concerned to employ as much of its existing manufacturing resource as is possible in order to produce a product quickly and efficiently, then the main additional constraints that have to be taken into account by the designer will arise out of the control and manufacturing activities.

Under-constrained, scheming-centred: Should it be the company's objective to develop a range of integrated furnishing products to meet the needs of a diverse commercial market, then the major design activity will be concentrated in the scheming stage. It will be the attention paid to the refinement and interrelationship of design details that will govern whether the products succeed or not.

Under-constrained, concept-centred: Finally there is always a market for the exotic or unique product. Here it is the original idea that takes precedence over all other considerations, and the design process thus centres on the concept stage. The chair designer may set out to develop a concept in order to achieve a particular effect, style or posture and may do this without any regard to how the resulting chair is to be manufactured or assembled. Companies working in this way may choose not to have a full manufacturing facility of their own, preferring to subcontract in order to preserve a truly flexible approach.

This brief outline of the wide variety of approaches that may be adopted in the case of a simple product such as a chair should make it clear why it is that the design process often appears to defy all efforts to describe or define it. Whilst common elements can be found, the process can only be directly understood in relationship to its effectiveness in creating a particular class of product. When considering changes in processes or techniques, it is always necessary to remember that the nature of design is determined by the nature of the product itself.

MODULE 1.3 TRADITIONAL DRAWING PRACTICES

Contemporary drawing office procedures have evolved over many years in order to satisfy the needs of the engineering industry. For a very long time the engineering drawing was the only accepted means of conveying to the workshop information about a component and the instructions as to how it should be manufactured. As companies began to exchange engineering drawings with each other it became necessary for them to agree on a common, unambiguous means of representing the geometric data containing in the drawings. Various standards thus emerged.

COMMUNICATION OF ENGINEERING INFORMATION

It must always be remembered that these standards are simply an aid to the communication of engineering information, the engineering drawing is not an art form in its own right. Unfortunately, too many engineers have been taught to behave as if the drawing were the end product of the design process. As a result, while the clarity of the representation is, or course, of paramount importance, far too much emphasis is placed on the style of the drawing rather than on how efficiently and effectively it conveys the necessary engineering information.

The purpose of a drawing office code of practice, and of the engineering drawing in particular, is to ensure that all the information needed to produce a functioning article from the design scheme is communicated clearly and unambiguously. The drawing does not explicitly convey the description of the function of an object; instead it provides the geometric details of a single chosen embodiment of the designer's original concept. None of the ideas behind the design are directly presented; if it is necessary to discover the purpose or role of the object, this will have to be deduced from its features and the way in which they are related to each other or to features of other objects.

DRAWING REPRESENTATION BY DRAFTING AND CAD

An engineering drawing is thus only a semi-pictorial representation of an object and is not intended to convey an exact visual impression. The views themselves are a series of two-dimensional representations of a three-dimensional object. How these views are laid out upon the drawing sheet is dictated by convention and various standards. Classically, the procedures used have been taught as a set of rules that must be complied with in order to produce such layouts as first angle, third angle and auxilary projections (see Module 5.1).

Today, with the advent of computer-aided design, it has become possible to think of these rules in a rather different way and look upon them as procedures whereby two-dimensional views can be derived from a three-dimensional model.

Traditionally, the draftsman has relied heavily upon a three-view construction (front, side and plan) in order to provide the clearest interpretation of an object which exists only in his mind. With CAD it is the model (often three-dimensional) that provides the *true* representation of the object and the views, from any chosen position, can be selected to provide the best display of those features which are relevant to any particular purpose. The role of the engineering drawing is thus changing. No longer is it the only interpretation of the object and all the information necessary for the manufacture of that object need no longer be reproduced upon a single drawing.

A REFERENCE FOR ENGINEERING INFORMATION

In the past the engineering drawing was seen as the sole reference source for the majority of engineering information. The CAD database has now taken over that role, and the drawing, in its various forms, has become the means of communicating discrete 'parcels' of that information to the user. A complete drawing file can be partitioned or 'layered' to allow the data appropriate to a particular activity or process to be easily accessed and plotted. The drawing can be reconfigured (whilst the database model remains unchanged) - perhaps to give coordinate geometry and values for prescribed settings and machine operations, to show the separate description of features and relationships for assembly operations or to provide reference coordinates and offsets to be used in the inspection of the finished object.

Changing Role of the Drawing within CAD

While the majority of CAD users are still producing drawings to the conventional specification, the few who have begun to explore the wider possibilities of CAD have reported spectacular productivity gains.

To take just one example. A company used the layering facility of its CAD system to show the state of progress at each stage of a lengthy reorganization of one of its factories. By selecting one set of layers that were continuously visible, and combining these with other layers that could be added to or subtracted from the image as required, it was possible to display a plan of the factory that showed the state of the floorspace at every stage of the planned reconstruction. Not only was the screen image easier to read than a traditional drawing, because it contained only the information currently needed, but it also allowed major errors in planning to be detected at an early stage. By displaying the sequence of states the site would pass through it was easy to establish the appropriate time for the removal of a structure or a machine and the use of this technique is reported to have allowed changes to be made to the schedule which ensured that equipment could be manipulated into position while still leaving the majority of the plant active and productive.

With the development of further links between automated design and manufacturing systems, many have argued that the engineering drawing will become a thing of the past. This will not be the case. Its form will change but it will continue to perform an important role. No matter how intergrated manufacturing systems become, it will still be necessary for people to be able to understand what is being built, how it goes together and how it works. Without this information they can have no confidence in the processes they are operating or managing.

The Engineering Database

With the role of the engineering drawing not finally settled, it is hardly surprising to find that the form and role of the CAD database are also matters for dispute. If the sole purpose of the database is to allow traditional drawings to be produced, then its usage will be restricted. But if it is considered as a product modelling facility, it can and should be used to monitor and control all downstream activities, including manufacture, assembly and inspection processes. The information contained in the database can still be extracted and used to construct engineering drawings which show particular aspects or features of the design. But the same data can also be fed directly into numerically controlled machine tools and coordinate measuring machines.

To sum up, the authority of the engineering drawing as the sole source of geometric and engineering information is in the process of being usurped by the CAD database, which will then provide the basis for the fully integrated design and manufacturing systems of tomorrow. But this does not mean that the day of the 'paperless' design office is at hand.

MODULE 1.4 THE DEVELOPMENT OF THE CAD INDUSTRY

The creation of a new technology and an industry to support it is dependent upon the emergence of suitable tools. But these tools will not be accepted into common use unless it has become clear that existing tools are no longer adequate. Both the new techniques and the desire for change must be present in order to catalyse an industry into the next stage development. The origins of any technological innovation such as CAD are thus murky, to say the least. Often the trigger is furnished by a number of separate events which are the result, not of serendipity, but of the work of individual researchers who are unknown to each other but who share the same broad view of the way in which a technology should advance. It was this way with the development of what we now call the CAD/CAM industry.

GRAPHICAL DISPLAYS

By the 1960s the development of the high speed electronic computer had reached a stage at which the automatic processing of vast amounts of numerical information had become commonplace and the plotting of the resulting outputs was becoming a chore. This led to the emergence of the cathode ray display terminal (or, more accurately, its re-emergence, since it had been one of the first output devices ever used on computers). The use of the CRT allowed pairs of variables to be plotted directly upon the screen, and techniques were rapidly evolved that allowed text to be added, together with predefined geometric symbols or shapes that were stored within the computer's memory. Soon, libraries of graphical software procedures appeared, providing a range of routines for drawing the common geometric entities (such as points, lines, arcs and circles).

INTERACTIVE DRAWING

The next logical step, the appearance of an integrated and interactive package, soon followed. The first demonstrable system is usually claimed to have been that produced by Ivan Sutherland at MIT in 1962 and called SKETCHPAD. This initially provided a two-dimensional environment in which a drawing could be produced. Later, in collaboration with Steven Coons, Sutherland extended his system to provide full three-dimensional images.

STYLED SURFACES

During that same period a team at the General Motors Research Laboratories was developing a graphics-based system which not only allowed styled shapes to be viewed on a display screen but also made it possible for models to be machined directly from the numerical data. The link between CAD and CAM was thus in existence from the earliest days. This led, in 1964, to the construction of the first true CAD/CAM system called DAC1 (Designed Augmented by Computer). One of the main researchers involved, Pat Hanratty, subsequently formed MCS Ltd, and produced the ADAM software package which was, in turn, to provide the basis of many commercial systems. ADAM was also the forerunner of a more advanced package, called AD2000, which developed into another range of commercial systems which are still the basis of much of the current CAD/CAM industry. It is only now, with the emergence of the latest generation of modelling techniques and hardware systems, that the link with these original programs is being severed.

DESIGN-BASED ORIGINS OF CAD/CAM SYSTEMS

The embryonic CAD/CAM companies which were established to exploit these technological developments set out to attack markets in a variety of sectors in the engineering industry. The various companies also chose to tackle differing aspects of the design process; some initially concentrating on the analysis aspect, some on scheming (drawing) and others on manufacture (see Module 1.2). Moreover, as these early systems developed, they tended to be tailored to suit the needs of a particular industry (usually mechanical or electrical engineering, architecture or cartography). It was only later that the companies which survived began to develop general purpose systems.

Analysis-centred systems found their main application in large-scale manufacturing businesses, such as the aircraft and motor industries. Companies in these areas rapidly absorbed and continued to develop the new techniques. Their primary need was for systems that could generate complex free-form surfaces to meet the conflicting requirements of aerodynamics, styling and structural strength. These over-constrained design problems led naturally to the development of highly sophisticated analysis-based programs. These industries had problems that had to be solved, no matter what the cost, and this meant that they were prepared to invest in what were, at the time, large and expensive computing facilities.

Drafting-centred systems became popular as a result of the advent of the mini-computer, which allowed CAD techniques to be applied in a cost-effective way to the area of design drafting. This provided the basis for the scheming systems. A range of products directed toward satisfying the needs of the engineering draftsman emerged. Initially they simply provided 'electronic drawing boards' which were rapidly enhanced to handle model manipulation, parametrics and pre-programmed graphics activities. Such systems, as with the original SKETCHPAD, soon offered three dimensional capabilities in a wire frame mode.

There thus emerged a range of products, many of which still exist today, covering various needs of the drafting aspects of design work. Some set out simply to provide a computer-based drawing facility. These contained routines to allow any engineering drawing to be constructed through the laying down, manipulation and modification of the normal drawing 'primitives' such as points, lines, arcs and circles; later developments added free-form curves (see Modules 3.1 to 3.6). The effectiveness of these systems lay in their ability not only to produce drawings faster, with fewer errors, but also to operate to, or change between, different drawing standards. The style of dimensions and the drawing layout could be set to conform either to an accepted national standard or to a company's in-house practices.

The main benefit of drafting systems was soon recognised to be the relative ease with which parts could be constructed: by using standard or library 'macros', by extracting relevant portions of an existing design or by the processes of copying or mirroring a single feature.

THREE-DIMENSIONAL MODELLING

Whilst drafting-centred systems could satisfy the limited requirements involved in producing engineering drawings, it was clear that there would be many advantages to be gained by having a system that allowed the design to be generated in three dimensions. Some attempts were made to provide a three-dimensional image by combining a pair of two-dimensional views, but true three-dimensional modelling first appeared in the form of the wire-frame. Here the edges of surfaces or planes are represented by lines and curves in space which provide a three-dimensional illusion of the object. The drawback is that such a representation includes all the 'wires' necessary to construct the edge lines of every surface or feature – even those normally hidden from view. As a rresult the image is often confusing and difficult to interpret. Hidden-line removal can be provided to a limited extent, but this involves making some assumptions regarding the nature of the non-existent surfaces between the frames.

Although wire-frame modellers are provided on the majority of 'turnkey' systems (integrated sets of hardware and software supplied by a single vendor), they have not been fully utilized by industry. This is partly due to the difficulties of interpreting wire-frame images and also because of the changes in design technique that their use necessitates. Many industrial users are only now beginning to encourage the use of full three-dimensional modelling in their design and drawing offices and the major benefits that can result from the evaluation and visualization of complete three-dimensional models during the design stage have yet to be achieved in many cases.

THE MANUFACTURING INTERFACE

Bringing CAD techniques to bear upon the next aspect of design, the manufacturing stage, was initially a matter of providing a set of graphical instructions that could duplicate the manual processes involved in writing instructions for existing numerically controlled machine tools. The first step was to generate toolpaths as geometric constructions within the CAD system; these could then be passed to a post-processor to be turned into machine-readable instructions that would generate the desired motions. Additional information such as cutter speeds and feed rates, was also edited into the instruction file.

Such toolpath descriptions were at first limited to simple two-dimensional routing or turning operations. The next, obvious, step was to combine a series of two-dimensional cutter paths, each set at a different depth, in order to machine a contoured surface (a procedure normally described as a $2^1/2$ -axis operation).

The development of a full five-axis surface operation (to cope with a truly free-form surface with the tool being controlled to align with the surface normal) required a the CAD system that was capable of providing a mathematical definition of a surface rather than merely its boundaries or edge-lines as in the case of a wire-frame modeller. This requirement, together with the need to generate the intersection curves of penetrating objects, led to the development of a range of surface modelling techniques.

Surface modelling can be seen as a natural extension of the wire-frame techniques, since it is based on surface equations that are derived from knowledge of the mathematical form of boundary curves and their derivatives. The contained surface can either have continuity of curvature with the adjacent ones or can be described as a series of discrete patches.

CURRENT SYSTEM AVAILABILITY

Systems that integrate wire-frame and surface modelling have been available commercially for a number of years, but they have yet to be widely used on industrial problems. Despite this, many vendors are now offering solid modelling systems which will bring an even greater capability to the design office. Again, this is an advance that will not be fully utilized until design office practices and skills have been developed to take advantage of it.

While the vast majority of the engineering industry is still attempting to come to terms with the current level of CAD technology, the research and development continues. In recent years the microcomputer has matured from the 'games machine' to become a stand-alone industrial CAD workstation or a part of a networked system. New and advanced solid-modelling techniques are appearing and expert or knowledge-based systems are being developed for CAD applications.

New developments in both hardware and software will continue to appear at an ever-increasing rate. Since this non-stop technological advance is itself driven by the need to produce a creative design environment upon a computer, it will only begin to level out as that goal is approached. We have already defined the design process (see Module 1.2) as one which involves the ability to handle complex interrelationships in a bounded but ill-defined problem-solving situation. Much further progress will be required before CAD systems are fully equal to this challenge.

EXERCISES

1. Take a product such as a bicycle lamp and identify the main functional features and their interrelationships. (Hint: in some lamps, the geometric design is controlled by direct associativity between the lamp bulb and the battery contacts.) Compare variations in product design. Also identify the limiting geometric parameters associated with each functional feature. (Hint: the shape of the reflector, for example, is determined by the desired light pattern; that of the housing by the battery configuration.)

2. Identify the functions that need to be provided in a power-winder attachment for a camera. Show that the overall volumes over a range of possible design solutions that can provide specified power and duration are very similar. (Hint: consider the relationship between the sizes of DC motors with differing torque characteristics and the gearing needed for a given output torque.)

3. Consider the design activities necessary in the development of the following items and indicate where the main effort is likely to be centred.

    (i)   a fork-lift truck
   (ii)   an aircraft
  (iii)   cutlery
  (iv)   a domestic washing machine
   (v)   a flower pot
  (vi)   a television set
 (vii)   an underground train

(Hint: select specific product examples to aid the consideration. This is helpful because most items can be derived in more than one way depending on whether, for example, they are directed towards specific or mass markets.)

4. Indicate how CAD could help in the development of the following products:

    (i)   a mould for a plastic bottle
   (ii)   a car body panel
  (iii)   a television transmitter mast
  (iv)   a printed circuit board
   (v)   a drilling jig
  (vi)   a folded metal box.

5. Suggest how the development of two-dimensional drafting systems has influenced changes taking place in the drawing office. Indicate also what role the engineering drawing now plays within the design process.

6. Describe how developments in computer technology have been reflected in advances in CAD.

7. Sketch a connecting rod for an internal combustion engine. By considering the functions it has to perform, indicate what parameters affect its main geometric dimensions.

8. Discuss how conflicts might be resolved within an over-constrained design problem. Suggest how the trade-off between parameters such as strength and weight may vary depending upon an assessment of the risks of failure and of the necessity for good performance over a long period.

SECTION 2
SYSTEM CONFIGURATIONS

The CAD industry has advanced hand-in-hand with the computer industry upon which it depends; it has benefitted from advances in computer technology and has also, in its turn, stimulated many of them. It is therefore not surprising to find that in CAD, as in computer design, a whole range of new developments are in prospect.

In the early days of CAD the power of computers was limited, they were very expensive and their modelling procedures were relatively crude. The large mainframe, which was the only type of computer available at the time, was viewed as a major company resource that had to be made available centrally in order to support the whole organization. It was required to fulfil many different roles and to support a number of users simultaneously. Contemporary thinking about the role of computer systems thus ran counter to the needs of the embryonic CAD/CAM industry. If a company invested in a CAD system it suddenly found that its computer, which had been designed to provide multi-user access via alpha-numeric terminals operating upon relatively small programs, was required to support a process that required large amounts of memory space, employed large programs and demanded substantial processing power. At best, it was necessary to restrict the use of CAD to a very limited number of graphics terminals. Design activities could not be performed upon existing systems without degrading their performance to a level at which all other tasks almost ceased to function.

The introduction of CAD/CAM and its ultimate acceptance as an industrial design and manufacturing tool was thus dependent as much upon the timely arrival of the mini-computer as it was upon the development of the graphics workstation. Even now it is still these two elements that are considered best to characterise a CAD/CAM system. All units possess a computational facility to handle and manipulate the geometric data, and all employ a graphics screen to display the results. Within this framework a large and growing number of variations have been developed, using differing approaches to satisfy the needs of a range of markets.

Eventually many of these alternatives may disappear or merge through common development objectives, but the needs of various specialist users will ensure that a reasonable range of options remains. The main thrust of current CAD/CAM development is in the direction of powerful, intelligent stand-alone workstations that can communicate with each other and be managed via a network. Such a system combines the advantages of concentrating the much-needed computational power at the workstation, where it is required, while allowing advice to be sought, information to be exchanged and operations to be controlled and managed through the network. It is systems of this kind that will enable CAD to realize its full potential in the context of the design process.

MODULE 2.1 WORKSTATIONS

Over the past decade the workstation has evolved from being a simple graphics display terminal into an input device of considerable complexity that can handle many of the CAD functions locally. It is composed of two major elements; a means of entering graphics commands and a screen upon which the results are displayed.

VECTOR DISPLAY SCREENS

Early systems relied upon vector, or stroke-writing, techniques using an adapted cathode ray tube. Vector-writing, which involves steering the electron beam across the screen (figure 2.1), produces a line or image of high quality.

But it does suffer from three major drawbacks that were to influence the way in which these early devices developed. Firstly, since the drawn image is produced by tracing the beam through all the chosen points, so causing them to fluoresce, it is in practice only possible to display one colour and a limited range of tones. Secondly, simply maintaining an image on the display screen required a considerable amount of computer memory to be devoted to storing

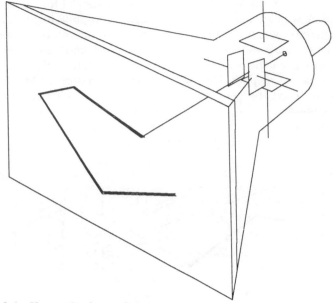

Figure 2.1   Vector display technique.

the tracking data (a very important consideration at a time when core memory was extremely costly). The final problem became evident as the amount of data to be displayed increased. Since the frequency with which the picture could be redrawn was dependent on the time taken to write each individual line segment, the speed of redraw dropped as the amount of data increased. Eventually the image developed a perceptible flicker which became distressing to the operator.

The vector display technique does however have an advantage in that changes in the displayed image or the position of a cursor can be rapidly and automatically updated on the next display cycle. This is particularly useful when animation or kinematics are being attempted. Vector display is therefore still widely used today in specialist applications and its limitations have been reduced by the vast increase in available computational power and advances in beam deflection technology. Although it is still unable to provide colour block-filling, some attempts have been made to produce changes in line colour by exciting differing fluorescent layers on the tube. However, this work remains very much a research tool.

DIRECT-VIEW STORAGE SCREENS

The first major development of the vector system resulted in the direct-view storage tube (DVST). In this system a main beam is employed to write the image on to the screen as in a vector display (figure 2.2), but the screen is coated with a special long-persistence phosphor which, once activated, continues to glow for as long as a second 'flood' gun bombards it. The image will therefore remain upon the screen until the flood gun is turned off and this relieves the computer from the necessity of storing the data and continuously redrawing the display. The CAD system computer is thus freed to concentrate on the generation and analysis of more complex geometric models.

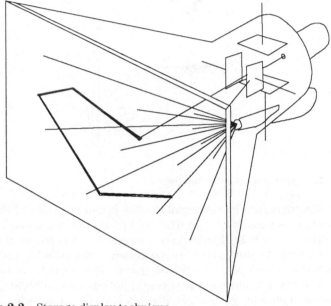

Figure 2.2 Storage display technique.

The main disadvantage of this technique results from the semipermanent nature of the image shown on the screen. Moves or changes cannot be easily erased; a moving line or cursor, for example, will leave behind a trail of images and these confuse the display. Objects can only be removed completely from the view by clearing the display and 'repainting' the whole screen. For large and complex views this process can be very time-consuming. This problem, together with the greater availability of computer memory has made other techniques increasingly attractive and has caused the storage screen gradually to go out of favour.

Raster Display Screens

An entirely different approach to vector construction has been use of the raster display systems (figure 2.3) already employed extensively in television broadcasting.

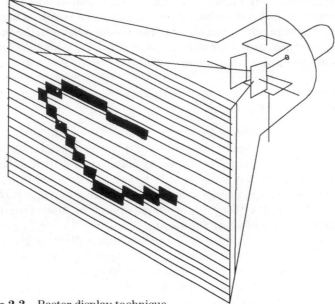

Figure 2.3 Raster display technique.

Here the screen is divided into a large number of point or 'pixels' that are assigned values according to whether they are to be illuminated and with which colour. The state of each pixel is held in a map within the computer memory (or within memory associated with the graphics device). In order to display a horizontal or vertical line, for instance, a row or column of pixels is switched over from displaying the background state to that chosen for the line. The image is then created and maintained by the computer which continuously scans through the pixel map held within its memory.

The most obvious disadvantage of a raster display is that a diagonal line is broken into pixel-sized dots or small line segments that, when displayed, produce a staircase effect referred to as 'jaggies'.

Although the raster system can be used to display colours, tones and block-fills, the quality of the image will depend upon the number of pixels that can be stored and displayed. Initially this limitation was based on the size of the available memory but the critical factor has now become the screen display technology. Screens with densities of up to 1000 by 1000 pixels are available commercially and provide displays of fairly good quality, though with jaggies still visible. Higher resolution systems can be produced, but advances in technology will be required before a significant increase in quality can be achieved at an affordable price.

The technical problems associated with colour displays centre around the need for separate guns which fire though holes in a masking screen to hit and fluoresce the three primary-coloured phosphors that, in combination, generate the true coloured image. As the number of pixels increases the density of the holes in the masking screen must also increase until it reaches a point at which the material is no longer strong or rigid enough to maintain itself in the right position.

SPECIAL-PURPOSE FACILITIES

Whilst the raster system is now used in most commercial turnkey systems, vector terminals are still used in some special-purpose and advanced applications. The latter usually employ their own local computer (which may be relatively large) to drive and control the whole display process. Such systems are able to zoom in and rotate a large and complex three-dimensional object in almost real time, generating effects of startling realism.

Other systems are under development that will allow even better quality displays of the modelled objects. Many flat screen techniques have been proposed, with the plasma screen being the first to emerge commercially. All of these developments indicate that the limits of the technology have not been reached and that there are still many changes yet to come.

Figure 2.4 General view of a CAD workstation.

CURSOR KEY INPUT DEVICES

Although all CAD workstations incorporate the traditional alph-numeric keyboards, familiar from a hundred and one other computer applications, it is also usual for this input device to be supplemented by another which allows the user to communicate directly with the graphics database and screen (figure 2.4). The simplest devices are those which allow a cursor to be moved around the screen in order to indicate the position at which a specific graphical entry is to be made. Advance systems may provide an array of such devices specifically designed for CAD/CAM activities, and the operator may only rarely have recourse to the keyboard.

In some cases the keyboard itself incropates additional keypads that are used to control the movement of the cursor. Many keypads will only allow the cursor to be moved in the four primary directions (up, down, left and right); others provide four additional keys that move the cursor along the diagonals as well.

DIRECT CURSOR CONTROL DEVICES

Other systems for controlling the movement of a screen cursor involve the use of thumb wheels, joysticks, tracker balls, pucks and mice. Of these, the mouse is the most recent innovation and it has been claimed that it is an entirely new form of input device. In principle, however, the mouse is simply an inverted tracker ball, with the ball being rotated as the body of the mouse is pushed across a surface rather than being turned by hand. An alternative form of mouse incorporates photoelectric devices that detect its movements as it traverses the surface of a tablet which has a grid printed on it.

Thumb wheels are usually grouped in pairs to allow the cursor to be positioned by independent movements along the horizontal and vertical axes. The joystick and tracker ball have the advantage that they can be used to move the cursor in any direction. In the case of the joystick an additional refinement may enable the user to indicate the speed at which the cursor is to be moved – the more the stick is displaced from its central, 'rest', position, the faster the cursor will travel in the direction indicated. Some joysticks have been further enhanced by including sensors that detect rotation of the stick, or even axial motion, this allows the user to convey additional instruc-tions controlling activities such as zooming and the selection of rotation angles. Such systems, which also incorporate 'firing' buttons mounted on the stick itself or the base, can enable the user to convey a very large number of CAD commands.

DIRECT POINTING DEVICES

All devices discussed above provide a means of manipulating a screen cursor. The pen, in its various forms, and the combination of puck and graphics tablet are, in contrast, devices which enable the user to 'point' directly at any spot on the display screen. A puck is a hand-held device with a cross-hair 'sight' that is used in conjunction with a tablet. By placing the puck on the tablet and pressing a firing button, the user indicates that he is 'pointing at' the corresponding position on the screen.

Two main types of pen systems exist: one, like the puck, is used to indicate points upon a tablet, whilst the other, the light pen, works directly upon the screen. The action of pointing at an object upon the screen is undoubtly very natural, but does mean that the object is obscured by the pen and user's hand and accurate selection is therefore difficult. This problem, together with the tiring effect of continuously raising the pen up to the screen, has caused the light pen to lose much of its former popularity.

TABLET AND MENU INPUT PROCEDURES

Most manufacturers of turnkey CAD systems have, up till now, favoured the use of the pen and tablet arrangement. There are, however, signs that the mouse is now catching up in popularity. The advantage of the tablet-based system is that the pen can be used not only to indicate points upon the screen, but also to select options from a menu that is in corporated into the tablet itself. In this mode of operation regions of the tablet are divided up into a grid of sub-areas, or 'keys', which can be 'pressed' with the pen.

This system is very flexible and allows a very large number of differing commands to be communicated easily and quickly. The system can also be tailored to a specific type of operation or application; alternative menus can be provided to go with different programs and it is also possible for operators to devise their own personal or preferred layouts. Many vendors provide sets of complete menus or partial overlays to accommodate the most common areas of application.

Some systems are now available that allow menus and icons to be constructed on selected areas of the screen. These may be used in conjunction with a tablet and pen or a mouse. The problems arising from the limited amount of screen space available (clearly, most is needed for graphics), have been overcome by the development of on-screen menus which can be 'pulled down' or overlaid on the screen area and which disappear when not required. Menus are arranged hierarchically so that the operator can rapidly drop down through a sequence of responses in order to converge upon a chosen command.

CHOICE OF INPUT DEVICE

The choice of input device is to some extent determined by the type of screen technology employed and the means whereby the CAD commands are selected. Limitations in screen resolution and size, together with restrictions on memory capacity, have, until recently, discouraged the use of on-screen menus. These restrictions are starting to be overcome and this means of command selection is proving to be popular. It is very versatile and allows a large number of commands to be easily grouped near to the graphics working area. This trend suggests that, at least for the immediate future, more CAD workstations will become available for which the major commands are entered by pen or mouse working from on-screen menus.

In the long term other forms of input techniques may become available. These could include comprehensive voice recognition, and sketch and character identification. Developments in the areas of expert systems, and in artificial intelligence generally, may also provide a means whereby natural language commands and free-hand sketches can be entered and correctly interpreted.

MODULE 2.2 COMPUTERS – MAINFRAMES TO MICROS

The heart of all CAD/CAM hardware is the central processing unit of a computer. The processor may, at one extreme, be no more than an apparently insignificant box placed next to the graphics screen; at the other, it may be a range of impressive-looking cabinets housed in a remote, air-conditioned room.

Historically, as has already been mentioned, large mainframe systems were designed as centralized facilities which would be 'all things to all men'. They were there to provide the computational resource for the whole company. Although they began by handling non-scientific activities, such as accounts and wages, computers soon found a role in supporting the design process. It was large design offices, particularly in the aircraft industry, that were to a large extent instrumental in the rapid development and application of analytical techniques such as finite elements. These new computational design tools soon started to impose upon the central facility an unexpectedly high work load for which it was not designed.

Time-sharing Facilities

At the time (we are now talking of the early 1960s) the normally diverse computer load of a company was best handled by the application of time-sharing techniques. In such centralized systems the computer is configured internally so as to appear as a seperate machine to each user, no matter how many people are actually using it at the particular time. The computer, due to the speed at which it can execute instructions and the way it is programmed to distribute its effort among its users, appears to be handling everyone's tasks simultaneously. This is, in reality, only an illusion, as soon becomes obvious when the number of users, or the computational demands they impose, increases; for the performance of the system rapidly degrades and users spend more and more time waiting for a response to their requests.

Time-sharing systems were usually configured on the assumption that, while a large number of users had to be supported, their individual requirements would only involve the running of relatively modest programs. But most of the activities associated with the generation of computergraphics and the handling of complex geometric models, and especially those involving design analysis techniques, are computationally very demanding. The installation of CAD software in a company-wide time-sharing system thus created a need to support individual users whose computational demands could easily swamp the whole system.

MINI-COMPUTER TURNKEY SYSTEMS

This problem was eventually resolved by the development of 16-bit and, later, 32-bit mini-computers. The relatively low cost of such machines allowed companies to dedicate a single computer to CAD/CAM activity, regardless of the other computational requirements of the business. The result was the arrival on the CAD scene of the turnkey system, a complete hardware and software configuration assembled, marketed and maintained by a single vendor. The turnkey approach not only made CAD/CAM readily accessible, even to those who were not computer experts, but also made it available at an affordable price.

Such systems were normally built around a single mini-computer that could support between two and eight individual workstations. The number of terminals that were in use at any time would depend upon the type of work being handled. Computationally demanding activities, such as three-dimensional modelling and analysis techniques, tended to degrade the system's performance so that the number of simultaneous users had to be restricted to only one or two. On the other hand, however, if the system was required to handle only relatively simple operations such as two-dimensional drafting, then a larger number of users could easily be accommodated.

The limitations on the amounts of computer memory available in the early mini-computers meant that CAD processes were heavily dependent upon accessing a hard disk. The majority of geometric data and program modules were held on the disk from which they could be retrieved as and when required and paged into the available memory. The response time of such systems was thus limited by the speed at which data could be transferred between the computer and the disk.

The original turnkey systems were, however, rapidly overtaken by the sudden explosion in memory capacity resulting from the replacement of magnetic core storage with silicon chips and the advent of the micro-computer. These developments had dramatic effects. The reported size of memory used in commercial CAD/CAM systems has increased something like a thousandfold in just over ten years; at the same time, the distinctions between different types of computer have become increasingly blurred.

MICRO-COMPUTER SYSTEMS

The micro-computer was first developed to provide cheap computing for the individual user. The original micros were 8-bit machines suitable primarily for playing computer games, but they rapidly evolved to address a much wider market. After some initial forays into many different applications (by an even greater number of companies, most of which disappeared as rapidly as they had emerged), this sector of the computer industry has settled down and split into roughly two market areas: that of games and home computing, and what has been called the 'personal computer' market. Distinctions in the latter area have again become blurred, for some personal micros have far out-stripped the original minis in terms of size and performance while many home micros have developed aspirations to be personal computers.

Today, CAD/CAM systems are available for use on micro-computers of every size and shape. The one lesson to be learned from this rapid growth in the range and scope of software, and in the variety of machines available to run it, is that we should no longer be surprised at what we find. A few years ago the use of micro-computers in CAD was strictly confined to the area of two-dimensional drafting, and although the number of micros in use in this area greatly exceeded the total number of turnkey systems installed, they were still dismissed by many as simply 'toys'.

But the advances, achieved over the past few years, both in memory size and computational techniques, now permit the personal computer to play a major role in the CAD/CAM industry. True, there is still a widely-held belief that it is limited to two-dimensional drafting, but the fact is that commercial packages now exist that allow micros to be used for three-dimensional wireframe modelling, surface modelling and even full solid modelling. Micros can now run programs that will produce a fully shaded pictorial image of a three-dimensional model, complete with reflections and transparency (figure 2.5). The whole range of design activities, from drafting, through full three-dimensional modelling, to the generation of numerical control data for manufacturing, can all be performed on a personal computer.

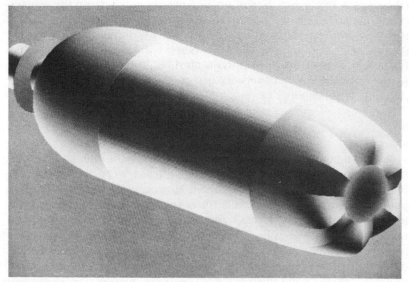

Figure 2.5   A fully shaded three-dimensional model.

THE WORKSTATION APPROACH

In the ill-defined no-man's land between the mini- and micro-systems there exists the workstation. This term is used to describe a stand-alone computational facility with the power of a mini-computer dedicated to the support of a single graphics display for a single user. Such systems are becoming more and more popular. When networked (see Module 2.5), they form systems which are likely to serve as models for the development of CAD/CAM well into the future.

FURTHER CHANGES IN COMPUTATIONAL FACILITIES

Advances in technology are already making systems installed perhaps as recently as five years ago seem crude. At that time many mini-computer-based turnkey systems were still being sold, principally upon the basis of their two-dimensional drawing capabilities with the purchasers showing little interest in the limited three-dimensional wire-frame modelling facilities that they offered. Currently, there is no sign that the pace of technological advance is slackening. Indeed, there is not even any indication that it is starting to level out. It would therefore be rash to predict what long-term developments may arise in such a volatile industry; it is even dangerous to try to forecast the next step with any degree of certainty!

Many improvements are likely to result from the combination of ever-increasing computer power and the growing availability of special custom-developed integrated circuits for CAD/CAM. Already, the means to accomplish the three-dimensional viewing transforms (together with the 'pan' and 'zoom' functions) has been removed from the CAD program and incorporated into the terminal hardware. This means that any geometric model held in the database can, once it has been transferred to the viewing terminal, be manipulated and presented from every viewpoint without any call being made on the computer itself.

Currently, a terminal's ability to perform such transformations may be limited to wire-frame interpretations; with other activities such as shading and hidden-line removal still requiring the services of the system computer. But one probably does not have to look very far into the future to foresee a time when the central computer will only be required to hold and manipulate those models that are actually in the course of construction, with all viewing processes being handled locally at the terminal. This reduction in computational activity, coupled with the steadily increasing availability of computational resources, will allow future systems to be applied to more creative and complex design activities. These may well involve the use of systems that have a greater understanding of the spatial relationships between objects; this would allow the correct functional requirements to be investigated at an early stage of design. Such systems could eventually allow the designer to 'discuss' his needs with the system and take 'advice' from a bank of engineering knowledge. The system would then interact creatively with the designer thoughout the complete design process, prompting him with advice and guidance while at the same time interpreting his responses in order to acquire a greater understanding of the task in hand so that its advice would be even more useful in the future.

In order that such CAD/CAM computers should come into existence in the not too distant future, it is only necessary that many of the advances currently being achieved in various areas of computer research and development should be brought together within a single system. The process control techniques employed in such a system would be based on methods developed in the fields of artificial intelligence and rule-based systems and would allow it to access a vast knowledge base, while the necessary increases in computational speed and performance would be achievable through parallel processing and other advanced array-handling techniques.

MODULE 2.3 OUTPUT DEVICES

The output of a CAD/CAM system can take several forms; but all of them can be thought of as a response to the user's request for data. Thus, the graphical display itself is the system's response to the operator's instruction that it constructs a particular view of a specified model. Similarly, the output of numerical values via the printer is the system's response to a command to list the values of a group of variables, and the transfer of a file of digital data to another computer for processing is yet another example.

PICTORIAL OUTPUT

Whilst any or all of these forms of output may be generated in the course of operating a CAD/CAM system, its most characteristic (and CAD-specific) outputs come in the form of pictorial images. Hard copies of construction details, models or drawings are taken by the designer at various points throughout the development process. These may be used to check on progress or may be referred to during any discussion with interested parties. Formal drawings are required for checking and approval upon completion of the proposed design. These may then be released to the manufacturing area either to provide the formal, documentary definition of the component geometry or as the authority for manufacturing staff to gain access to the geometric database in order that they may generate NC information directly from the computer model. Other drawings may be reproduced or duplicated in order to provide artwork for technical illustration, for example, or to be filed away (often in reduced form) as reference data for archival purposes.

The range of requirements for a CAD system to produce outputs in the shape of plots or drawings is thus very wide. Sometimes all that is needed is an image of low quality that can be produced quickly and easily at the graphics terminal. On other occasions a plot of very high quality and accuracy will be essential and the speed with which it can be produced and the whereabouts of the output device will be of secondary importance. Fortunately, a whole range of plotting devices are available, we describe the major ones below and give an indication as to their principal areas of usage.

TERMINAL PRINTERS AND PLOTTERS

Most turnkey systems provide some facility that enables the operator to produce a facsimile of the current display screen. This usually takes the form of a small plotter or printer into which the information used to create the screen display can be redirected. One such device, the thermal printer constructs its images by scanning the rastered lines across the paper substituting printed dots for the screen pixels. The resulting plot is thus a direct copy of the screen and does not represent the accuracy of the actual geometric information stored in the database. The paper plot may also include screen characters and prompts and will reproduce sloping lines and curves complete with jaggies, just as they appear on the screen.

The advantages of the thermal printer are that it is quick and cheap and one can easily be provided at each workstation, or, alternatively, one can be shared between a group of workstations. Its main role is to provide low quality pictorial information that can be used to check progress or to verify features while design work is actually in progress. These drawings can be filed as a record or referred to during discussions that take place away from the terminal.

The thermal printer can also be used to record textual information; it can, for example, print out a list of all commands given to the system and the responses received from it. This allows a sequence of complicated CAD commands to be recorded and retained for future reference so that the performance of the system can be analysed or its status at any given time can be checked. Records of this kind are useful for checking data and provide a means of communicating information between members of the same design team.

PEN PLOTTERS

The graphics hardcopy device most commonly used in conjunction with a CAD/CAM system is the pen plotter. Although pen plotters come in a wide variety of forms with several different configurations, all are designed on the same basic principle: by creating a relative motion between paper and an ink pen they create a complete drawing.

In some pen plotters the pen is attached to a carriage which moves back and forth along an arm which itself travels at right angles to its length; this arrangement thus allows the pen to traverse a sheet of paper along both its principal axes. Such flat-bed plotters are usually constructed so that the horizontal plotting data controls the movement of one element, say the pen carriage, while the vertical data controls the other. Although this may seem to be most obvious arrangement, others do exist, including articulated arms and even a fully steerable 'free-ranging' robot (or turtle) programmed to drag a pen across the paper – it is also programmed to check the paper first, to set its own origin and to untangle its trailing cables!

The next major category of plotters are those in which the pen is moved in only one direction while the paper moves in the other. The paper may be laid upon a flat, moving surface, passed over a drum or mounted upon a continuous belt.

Flat-bed, belt-mounted and large drum plotters can be used to create accurate plots using single sheets of paper or film. The flat-bed plotter is the easiest to set up for large sheets but it does take up a considerable amount of floor-space. The size of some such systems can be gauged by the fact that one is reported to be large enough to reproduce the complete side view of a truck full size. Drum plotters, on the other hand, can handle sheets up to A0 size and occupy relatively little space, but they do require more care to be taken when mounting the paper. To some extent the belt systems represent a compromise between these two extremes. They occupy no more floor space than a drum, but the fact that the flat section of the belt is almost vertical makes it easier to mount and set the paper.

Yet another variation of the drum plotter is a machine that allows paper to be wound on to the drum from a continuous roll and also allows the completed drawings to be wound onto a take up roll. Such systems are particularly useful when large numbers of drawings need to be plotted, since they can work through a whole set of drawings unattended, even overnight.

Plotting devices are normally used to generate reference or production drawings. Small plotters, up to A3 size, are often attached to a single workstation or personal computer for this purpose. Larger plots, up to A0, are usually provided by a central facility that is accessible to all terminals. This arrangement is dictated by the high cost of a large plotter, the relative infrequency with which it will be used in most circumstances and its need for constant supervision and maintenance.

SPEED LIMITATIONS

Pen-plotters can achieve a high degree of accuracy; some machines, for example, are able directly to plot the artwork necessary for the production of printed circuit boards. The main limitations are the speed at which they work and their inability to produce shaded images. As the drawing of each line or curve is a separate operation, and as there is a maximum speed at which the pen can be moved across the paper while still producing a clear plot, the plotting time increases in direct proportion to the complexity of the drawing. On some large and productive CAD/CAM systems it has been found that a single pen plotter may be unable to cope with the peak load and this can create unacceptable delays in the development process.

ELECTROSTATIC PLOTTERS

The electrostatic plotter has been developed in response to the need for a device that can produce a large and complex drawing of production quality rapidly and accurately. It is based upon a scanning process, as is the thermal printer, but, unlike the latter, it derives its data directly from the database rather than from the pixel map that generates the screen display. Essentially, the information taken from the computer database is fed into a computational facility built into the plotter which then generates the image. The system then proceeds to transfer the assembled image, pixel by pixel, onto the paper, using normal electrostatic copying processes. As the image is generated by laying down a large number of dots, electrostatic plotters are not restricted to line images and can in fact be used to produce coloured and shaded pictures.

Such plotters, while being very versatile and fast, are however relatively large and expensive. They are thus usually found as a central facility in large and specialized systems.

LASER SCANNING AND MICROFILMING

Whilst the speed of a pen-plotter is limited by the speed at which the ink flows, the speed of a photographic technique is determined by the virtually infinitisemal time it takes for the chemicals involved to react to the stimulating light. The drawing process can, therefore, be vastly accelerated by using a laser to 'draw' the image onto a sensitized sheet. Although such devices have not as yet found a large degree of acceptability in the production of normal engineering drawings, they have been applied to the generation of microfilm that is used extensively for archiving and for easy distribution of production drawings to remote sites. The principle is simple: a very fine laser beam is used to 'draw' the image directly onto a photographic film. The laser beam can be steered with a high degree of accuracy, even over a single frame of 35mm film and thus no further reduction process is required before the film is developed.

SUMMARY

The range of devices available for generating graphical outputs is, as we have seen, extensive and, between them, the various types of plotter can cater for many differing design activities. The choice of output device must be determined by the purpose for which it will be used and not simply by comparing cost against plotting accuracy. Employing a large and accurate machine that is difficult to operate, when all that is required is a quick plot of reasonable accuracy to check the progress of a design, is as bad as selecting a small and cheap flat-bed plotter to operate in an industrial situation where large quantities of accurate and complex drawings are being produced.

MODULE 2.4 DATA STORAGE

Within a design environment it is not only necessary to create engineering information but also to ensure that the data created can be stored in an orderly fashion and retrieved by the appropriate person as and when it is required. Information that has been generated on a CAD system will have differing status and be subject to different levels of authority at different times as the design process proceeds.

During the early stages many ideas are being explored. Some are easily rejected, others are of doubtful value, and a very few may offer a route to a possible solution. Any scheme is therefore likely to be provisional, at best, and will undoubtedly be subjected to many changes before its fate is finally decided.

Once the design has been completed, approved and accepted for manufacture, it is necessary to ensure that it takes on a more permanent status. The right to make changes must be restricted to ensure that unauthorised modifications do not take place once the part is in production. On the other hand, it is undesirable to 'freeze' the drawing or the computer model to such an extent that it cannot be revised or amended if major faults are discovered in the course of manufacture.

A rigorous but sympathetic set of control procedures must therefore be imposed to regulate and authorize access to geometric data, both while it is confined to the CAD/CAM system and after it has been issued.

MANAGEMENT OF DATA

It should be axiomatic that the geometric data representing a model should be filed and recorded in a sensible fashion. But a large number of currently available CAD/CAM systems are only just beginning to provide techniques for the proper management of their databases. Indeed, there has up until now been a strange and unexplained assumption that the data on a system would somehow organize itself – that the user could simply dump in all the data, as though the system were a dustbin, and somehow the computer would understand what was required and arrange it accordingly. Such an approach was never tolerated in a manual drawing office register, so why should it be accepted on a CAD system?

All the best of the well-established and highly effective drawing office practices should be transferred onto the computer. In particular there must be a recognition that with CAD, as with the drawing board, the designer needs to be able to create and experiment with a range of alternatives, most of which will not find a place in the final design. Those ideas that lead to the chosen solution may be incorporated in the main file, the remainder are usually destroyed. But even the discarded schemes may be relevant in that they reveal the problems that were found to be associated with various alternatives or the advantages that could result if certain changes were to be allowed in the specification.

The data storage and retrieval system that is selected must therefore tread a middle path between two extremes: one would involve retaining all information generated, no matter how irrelevant, and the other would involve discarding all details that did not form part of the solution finally adopted. The argument for keeping the background material is that it may be needed again in the future if an unforeseen problem arises with the part. But there is little point in this material being available if it is so unstructured that the relevant portions are difficult or impossible to retrieve.

There is therefore no substitute for putting in time and effort at the end of the design phase in order to make sure that those files that are to be kept are carefully selected and properly documented. All files should then be classified, under headings appropriate to the information they contain using the following categories.

MAIN DESIGN FILE

At the completion of any design task, there must be a main file which contains all the relevant design information for which approval is sought. It is this document that is accepted for production. As changes are made and approved it will be this file that will be modified and reissued. At any stage there can be only one approved production document whose status and record of changes are known. The main design file must fulfil that role, just as the master drawing did under the traditional drawing office procedure.

DIRECT SUPPORT FILES

An approved design may be based upon a particular arrangement of components and geometric constructions. Such information, whilst not important as manufacturing data, may be invaluable in establishing whether knock-on effects are likely to arise from changes made to other components or to the geometric relationships between components. If, during manufacture or service, problems are experienced with the part then the direct support information should be reviewed to see whether assumptions made at the design stage are still valid. Such files should be maintained and 'attached' to the main design file but not be made freely available beyond the design level.

ALTERNATIVE DESIGN FILES

In order to provide for the possibility that a problem may emerge with the chosen design, it is necessary to retain even those ideas that have been abandoned at an early stage. These schemes, together with the reasons why they were not pursued, should be held in such a manner that they can be reinvestigated at a later stage should it become necessary. The files containing these discarded schemes, direct support files, should only be available at the design level as they contain information that will be used to support or dismiss alternatives to the design contained in the main design file. They are therefore again attached to, but not released with, the main file.

PERSONAL FILES

During the process of designing a particular part the designer may generate constructions or derive solutions that are of general interest or which can be applied in a similar way to future designs. Each designer should therefore be encouraged to create a series of personal files that can eventually form part of a library of good design practice for the whole design office.

DESIGN APPROVAL

Once the design has been approved all design related files should be tidied up to leave the two main groups: the main design file and the attached files (both the direct support and the alternative designs). All other files should be deleted or transferred to the designer's personal directory.

How the data is stored and communicated beyond this approval stage will entirely depend upon the nature of the system. If the workstation is attached to a central computer or connected via a network, then the main facility will be responsible for the initial short-term storage and the longer-term archiving.

SHORT-TERM STORAGE

On a centralized system (and upon many personal computers), the short-term storage is handled by a fixed hard disk. come in various sizes and can usually hold between 10 and 300 megabytes of data. In the case of large disks, this can represent up to approximately 500 drawings, but this figure can vary greatly as it is obviously very dependent on the complexity of the drawings involved. Large multi-terminal systems will be serviced by a number of such disk units which usually have interchangeable disk packs. These disks can then be copied and back-up one onto another. Non-current data can be retrieved by requesting that a particular pack be remounted.

ARCHIVING AND RETRIEVAL

In principle, the hard disk offers both flexibility of retrieval and a secure means of safeguarding data against loss or damage. It is, however, an expensive way of holding old, archived material. Once a design has been frozen or a part has been replaced, it is only necessary to retain the information relating to it for historical or legal reasons. A strategy should therefore be laid down for determining when data is to be removed from the hard disk. This should not be done while the data is regularly being used, nor should it be delayed so that data is still on the disk years after it was last accessed. In finding a middle path between these two extremes, attention should focus on the status of the part rather than upon the frequency with which the data relating to it is retrieved. Removal of data from the system should be seen as indicating either that the part concerned is now in permanant use and will never require modification or that it has become redundant.

Once the decision to remove an item from the system is taken a secondary decision has also to be made as the whether it will ever be necessary to put it back on again. If the decision is that it should be recoverable then the obvious means of storage is on magnetic tape. Large reels can be used to store many drawing files but they do need to be handled, stored and maintained carefully. Periodicaly, the tapes need to be rewound, inspected and copied to stop the data from becoming corrupted as the tape ages.

Taped archives can therefore be costly to maintain if a very large number of files is involved, but they are fully recoverable. Naturally, as time passes, recovery may become less worthwhile, or indeed impractical, because of changes in design procedures or because the software and hardware of the CAD/CAM system has changed significantly in the meantime. In such cases the only practical means of recovering the data is to re-enter it by hand. It then becomes pointless to hold archives consisting of unreadable tapes. Before this stage is reached it is therefore sensible to decide which information will be maintained in a recoverable state (by translating it into the new system format) and which will only be recoverable by re-entry. Data in this final category is then best filed in an easily readable form, preferably as pictures or drawings. These can be taken off as master prints or as microfilms that can be stored in the normal manner (see Module 2.3).

One further storage medium exists for data files: the floppy disk. This is usually associated with the personal computer. But, just as it is quick and straightforward for anyone to take copies of 'floppies' on such machines, so, by the same token, the use of floppies makes it easy for any design procedures or authorisations to be abused. If a number of stand-alone personal computers are used in a design environment the taking of 'illegal' copies must be ruthlessly controlled in exactly the same way as unofficial prints are 'stamped on' within the traditional drawing office system. Officially approved files need to be held (and guarded) by a specified person who is responsible for the issuing of all copies or prints.

CONTROL OF DATA

All storage and distribution systems are subject to abuse, if only because the design process requires that the information should be freely available and easily changeable at one stage, and restricted and consistent at a later stage. The procedures adopted must thus be able to handle these conflicting requirements. In the end there is no substitute for the well-tried and traditional practices of registering, issuing and recording the status of all design files thoughout the entire development cycle.

MODULE 2.5 NETWORKED SYSTEMS

The early CAD/CAM systems were all configured to have a number of terminals dependent upon a host computer (figure 2.6). This central computer not only provided the computational power for all the workstations but also housed the unified database which was accessible to all users (subject to certain safeguards). Additionally the host was responsible for all communication with other devices such as plotters, tape readers and other computers.

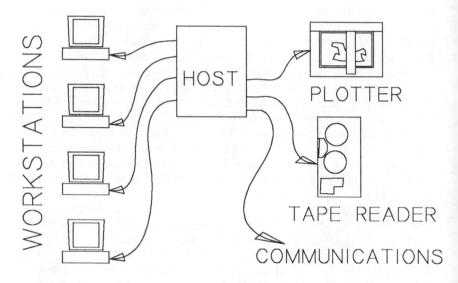

Figure 2.6   Host computer system supporting a number of workstations.

As the number of terminals increased and the individual computational activities became more demanding, the need to depend upon the host severely limited the response of the system. This led manufacturers of CAD systems to veer away from the centralized approach towards networks of stand-alone computers (often of the personal type) which are interconnected so that they can communicate directly with one another (figure 2.7).

INDEPENDENT WORKSTATIONS

One type of network is made up of a number of stand-alone workstations each of which is completely independent and reliant on no other unit for its computational power, memory or local storage. The network exists only to allow data to be exchanged between workstations. Large and expensive output devices, like the plotter and tape units, can then be coupled to a single terminal and accessed by all the other units through the network.

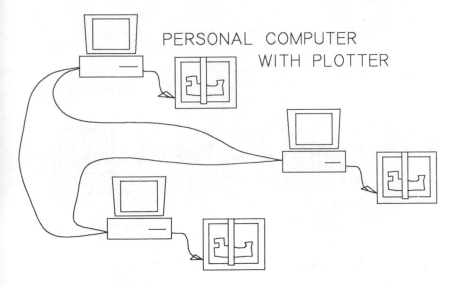

Figure 2.7 A network of personal computers.

Within this type of network, the design information is resident upon the unit on which it was first created. Any copies passed to another computer will result in a duplicate file being produced which may, after it has been modified on the receiving machine, result in multiple (and differing) versions being available on the network from different computers. Such a state of affairs clearly cannot be allowed to exist, so the network must be arranged to include file-handling procedures.

NETWORK FILE-HANDLING

The network management system will allow a file to be resident on only one machine, although it will permit it to be accessed and worked on from any of the workstations. The user thus has no knowledge of where in the network the file actually resides; for all operations are arranged to be 'transparent', making it appear as though all the processes are occurring locally.

As the complexity of the network increases it may be advantageous to attach special service computers, or file-servers, to perform and regulate the file-handling tasks (figure 2.8). These take over the primary role of storing, communicating and retrieving the files. They may also handle all communications to the plotters and tape units, as well as the links to other computer systems.

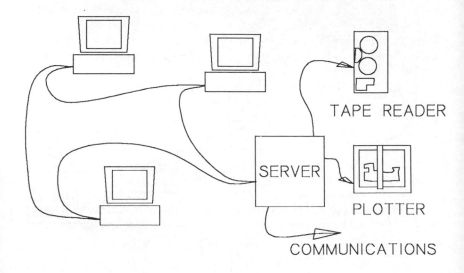

TAPE READER

SERVER

PLOTTER

COMMUNICATIONS

Figure 2.8   A file-server on a network.

NETWORKS REFLECTING THE DESIGN PROCESS

The further development of networking facilities will reflect aspects of the traditional manual design process. Information will be handled in differing ways at differing stages depending upon the level of definition that has been reached. The design will be accessed, manipulated and progressively enhanced by one set of experts, with one set of skills and responsibility and access will then be handed over to another set of people with a different kind of expertise. Thus, the designer will be responsible for the development of the scheme; the draftsmen for the geometric definition; the analyst for its modelling and verification; the part-programmer for generating the tool paths and other machining instructions. It would, in fact, require only a comparatively modest extension of the current networking technology to produce a system which provides facilities of this kind. The result would be the integration of the control and management processes into a genuine CAD/CAM environment.

In such an environment the management procedures would be used to direct activities, such as design, to the appropriate terminals together with the relevant briefing material. These designing terminals would then be responsible for controlling and developing the scheme to a state that was acceptable to the management process with the help of their own group knowledge. When the acceptable state was reached the relevant information could be extracted from the designing units and instruction sent to the detailing terminals whose task it would be to generate a final geometric arrangement. Detailed items, upon reaching an acceptable state of completion, would then in their turn be passed on to the analysis unit for verification. These procedures, whilst still conforming to the design scheme, also comply with the manufacturing constraints.

Such an intelligent networking arrangement puts the expertise back into the hands of the appropriate staff and does not require everyone to become an expert in all areas of design through to manufacture. The constraints and requirement of all aspects of the design are then communicated and maintained by the management process. This keeps track of any violations and resolves conflicts as they arise. The resulting system is then able to respond to all the conflicting requirements of flexibility, data consistency, design and manufacturing constraints whilst providing an environment which is both creative and capable of responding rapidly right through to the production stage.

EXERCISES

1. Outline the main advantages and disadvantages of the raster display system over those of the vector display system. Identify those applications in which the vector approach is preferred.
2. The CAD user interface can be designed to employ a range of input devices. Describe the functions that can be performed using the following devices and explain how each interacts with the CAD software:

 thumb wheels
 mouse
 tablet and pen

3. Indicate how on-screen menus, arranged in a hierarchical form, can be used to help the operator select CAD commands in a commercial system.
4. Discuss the current and future roles to be played by the central processing unit (CPU) in an industrial CAD/CAM facility.
5. Indicate how the personal computer is likely to be employed within the design process in the future. Select the design and development activity for a particular product and describe how the individual processes could be handled by such a computer in an interactive manner.

6. The archiving of engineering information records is an important aspect of maintaining a company product database. Describe methods appropriate to the short- and the long-term storage of that information (with and without retrievability).

7. The production of engineering drawings on a CAD system can create a large workload. Show how the effects can be minimised by the selection of suitable equipment to fulfil a range of roles in the design office.

8. Suggest effective methods for regulating and constraining the creation of multiple records of engineering information that can be implemented on a network of CAD workstations.

9. Describe the input and output devices appropriate to a stand-alone, personal computer-based system used solely in the production of engineering drawings. How could this choice be varied if several machines are networked together?

10. The role of the workstation is to provide an interface between the designer and the model being created in the computer. Suggest how this could be changed by the development of a truly three-dimensional display and input system.

SECTION 3
ENTITY DESCRIPTIONS

At the heart of all CAD software systems is the programming that places the geometric information entered by the user into a database. Because a computer operates by processing numbers, this information must be encoded numerically. This is why the ideas of Cartesian coordinate geometry are important; they relate geometry to numbers.

When geometric manipulations are carried out by the machine in response to a user's commands, complex numerical procedures must be invoked to produce the desired results. Such procedures often will often seem 'obvious' if carried out manually – the location of the intersection of two lines, for example, apparently presents no problem to the human eye and brain. But nothing is obvious to the computer. It has to work through detailed solution algorithms and rely upon its speed of operation to ensure that the time spent is not frustrating for the user.

Part of the skill of designing CAD systems lies in finding a way in which the user can be given access to complex geometric procedures without being burdened with the need to have a detailed knowledge of all their intricacies. It is, however, a good idea for him to have some overall view of what is going on within the system. Things can go wrong; assumptions can be violated; and unexpected results can be produced. A user with some knowledge of the way the system works is in a better position to know just what operations can be carried out with certainty on any particular CAD system and can move easily from one system to another because he knows what is fundamental and what is incidental because it only relates to a particular computer implementation.

The earliest CAD systems were confined to two-dimensional geometry involving straight lines and circular arcs. Even today, cheap commercial computer-aided drafting packages are still available which can handle only these simple entities and they continue to provide a useful tool for companies whose products are all variations of a simple, basic geometric form (for example, office furniture). As CAD systems developed, it became possible to construct three-dimensional geometric models within the computer and the range of entities was extended to include more exotic geometric curves, such as conic sections, and also free-form curves which are particularly useful for aesthetic design.

The move into three dimensions soon led to the introduction of geometric entities representing surfaces in space. At their simplest, these include planar areas bounded by straight lines. More complex forms include ruled surfaces, cylinders, more general surfaces of revolution, and free-form surfaces, such as those used for car body panels. As the complexity of the geometry increases so too does the complexity of the routines used to manipulate it. The manipulations that are required include finding the lengths of arcs, the volumes enclosed by surfaces of revolution and the intersection of curves and surfaces.

In the following modules, we shall first look at the ways in which entities are defined and held within CAD systems. Their manipulations, especially the problem of finding intersections, are then discussed. The aim is to provide an overview of the way in which CAD systems hold and process geometric data.

MODULE 3.1 POINTS, LINES AND CIRCULAR ARCS

Computers are devices for dealing with numbers. They originally evolved as very rapid calculating machines and, although their applications are now very wide-ranging, the information they process is still essentially numerical. Even when numbers seem far removed from the application in question, as is, for example, the case with a word processor which handles mainly text, the information input to the system must be encoded in numerical form for the computer's own internal purposes. Thus the individual letters that make up a document held in a word processor have been converted into numbers by the machine, normally by use of the ASCII code.

CARTESIAN COORDINATES

The purpose of a computer-aided design system is to store, hold and manipulate geometric information. It is necessary to find a way of converting this into numerical data so that the computer can process it. It is here that use is made of the ideas of Cartesian coordinate geometry. If we are working in two dimensions, that is in a plane, we can encode the position of any point in the plane by means of two numbers. To do this two straight lines, called axes, are chosen, usually running at right angles to each other. The point where these axes intersect is called the origin. Any point can then be defined by two numbers that represent its distance from the origin in the directions of the two axes. The two axes are conventionally labelled x and y and the two distances used are referred to as the x- and y-coordinates of the point. A direction needs to be assigned to each axis and if the point being considered is on the same side of the origin as the direction of the axis, then the relevant coordinate value is taken to be positive; if not, it is negative. Some example of points and their Cartesian coordinate values are shown in figure 3.1

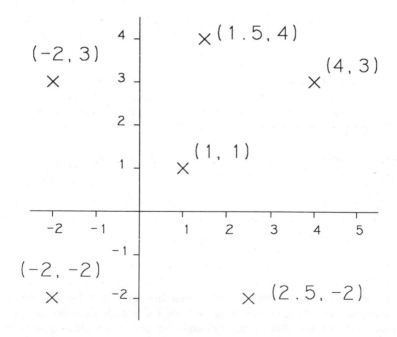

Figure 3.1 Cartesian coordinates.

The advantages of using CAD rather than the traditional drawing board become even clearer when the system has the ability to construct a three-dimensional computer model of the component being designed. It is true that, for reasons of simplicity, some commercial systems, particularly drafting systems, are limited to working in only two dimensions, but, from a theoretical point of view, there is not a great difference between using two or three dimensions and we shall therefore introduce three-dimensional coordinates at this stage. Three axes, each perpendicular to each other in space, are required. These are labelled x, y and z, and they intersect at a single origin. This enables us to define any point in space by three numbers which represent its distance from the origin in directions parallel to the axes. These numbers are the x-, y- and z-coordinates of the point. Figure 3.2 shows a typical point and how its coordinates are obtained.

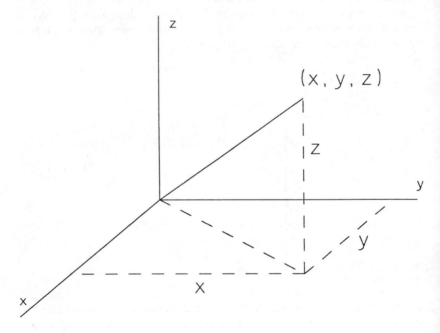

Figure 3.2   Three-dimensional coordinates.

NODES

The points in space whose positions are encoded by means of Cartesian coordinates are referred to as nodes and within the computer system a list of nodes can be stored as an array of sets of coordinates. For example, we can consider the rectangle shown in figure 3.3 to be composed of four nodes, namely its corners. If these were entered at the start of the node list the following would be obtained.

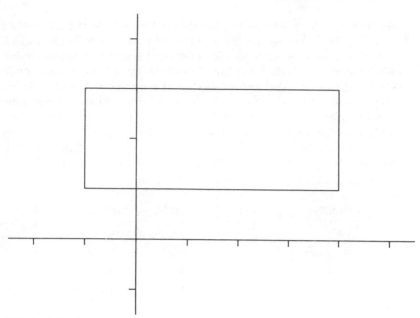

Figure 3.3   A two-dimensional rectangle.

Node	x-coordinate	y-coordinate	z-coordinate
1	−1.000	1.000	0.000
2	4.000	1.000	0.000
3	4.000	3.000	0.000
4	−1.000	3.000	0.000

Note that, since all the z-coordinates are set to zero, we have in fact described a purely two-dimensional object.

ENTITIES

The node list holds geometric information; it says where in space the nodes lie. However it does not tell us what to do with them. In order to connect the nodes together to form a rectangle we need extra information; we need to know about the 'topology' involved. For this an entity list is required. If the rectangle is to be described fully the following pairs of integer numbers will have to be held.

Node 1	Node 2
1	2
2	3
3	4
4	1

Each pair of nodes represents the end points of one of the four lines and these lines are regarded as the basic geometric entities that make up the rectangle. These pairs of nodes, together with the associated node list, thus describe the four entities which make up the model.

If all the entities are straight lines, two nodes are sufficient to define each of them. But if the model includes circular arcs three nodes will be needed for each, and four or more nodes maybe required for more complicated geometric entities. There are several ways of forming the entity list. The one we adopt here assumes that three nodes are involved in the majority of entities and the list consists of a collection of triples to each of which is added a code number that indicates the class of entity involved. If we use 2 as the code for a straight line, then the entity list for the rectangle looks like this:

Entity	Code	Node 1	Node 2	Node 3
1	2	1	2	0
2	2	2	3	0
3	2	3	4	0
4	2	4	1	0

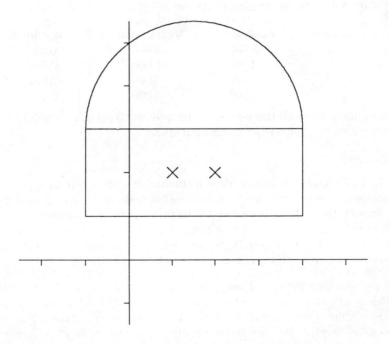

Figure 3.4   Two-dimensional geometry.

It is clear that, when dealing with straight lines, this form of the list wastes a certain amount of space by allowing space for an unnecessary third node. When other types of entity have to be dealt with, however, it has the advantage of making it relatively simple to gain access to the nodes for the various entity types allowed.

As already mentioned, three nodes are required to describe a circular arc. In the examples which follow, we take the first two nodes to be the ends of the arc and the third to be the centre, and arcs are taken to be drawn in the anticlockwise direction, starting at the first node. The code number 3 is used to indicate that the entity is indeed an arc.

We also introduce at this stage a third class of entity. This is a point. In the entity list it is represented by a single node and the code number used is 1. Thus, in some senses, a point is merely another word for a node. There are, however, advantages of uniformity to be gained by distinguishing the idea of a point from the idea of a node.

Node and Entity Lists

Using this three-part code, we can describe the two-dimensional shape shown in figure 3.4 by the following node and entity lists.

Node	x-coordinate	y-coordinate	z-coordinate
1	−1.000	1.000	0.000
2	4.000	1.000	0.000
3	4.000	3.000	0.000
4	−1.000	3.000	0.000
5	1.500	3.000	0.000
6	1.000	2.000	0.000
7	2.000	2.000	0.000

Entity	Code	Node 1	Node 2	Node 3
1	2	1	2	0
2	2	2	3	0
3	2	3	4	0
4	2	4	1	0
5	1	6	0	0
6	3	3	4	5
7	1	7	0	0

Commercial CAD systems may well hold more information in the entity list. Values such as the length of a line or arc, the angle a line makes with the x-axis, the radius of an arc, and so on, can all be held. These values are calculated when the entity is first constructed or modified and are then available for immediate use whenever they are required by subsequent operations. It is not, in fact, essential for the system to hold them, as they can always be calculated as and when required, but it will save time if they are needed often. 'Display attributes', such as the colour an entity is to be shown in or whether it is to be drawn using a solid, dashed or chain-dashed line, will also be held as part of the entity list.

MODULE 3.2 BASIC GEOMETRIC MANIPULATIONS

In this module we identify three main classes of geometric manipulation that can be performed on a CAD system: insertion, deletion and amendment.

INSERTION

In order to add a new entity to the entity lists it is necessary to establish which nodes comprise the entity. These may be existing nodes or may have to be newly defined. CAD systems use a variety of means for indicating nodal positions. In the most basic systems, the user may be invited to enter the actual coordinates via the keyboard. These coordinates are then inserted into an appropriate vacant space in the node list, thus creating a new node. Alternatively the user may form a new node by moving the screen cursor to the desired point on the screen. The accuracy of this method is of course limited. A refinement is to use the cursor to indicate an existing node by reference to the geometry already displayed on the screen. If, for example, the cursor is moved to a position near the end of an existing line, the end-point of that line can be taken as the required nodal position.

Depending upon the implementation of the CAD software, a new node may or may not be created in such cases. In the same way, it is possible to design the software so that the mid-point of an existing line or the centre of an existing arc can be selected as a new node. Once all the nodes required to define the entity being inserted have been defined, the entity itself can be created by adding the appropriate entry to the entity list. At this point the entity will usually be added to the display on the terminal screen.

DELETION

Deletion of an entity is straightforward. The user simply needs to tell the system which entity is to be deleted. This can be done by explicitly stating the number of the entity (if this is known) or by using the screen cursor to point to it. Many CAD systems will allow a 'mask' to be used. That is to say that the delete command takes the form of 'delete line' or 'delete arc' etc; the cursor is then positioned and the nearest piece of geometry of the required type is selected. Normally the entity that the system thinks has been chosen is highlighted in some way. If the user confirms that the choice is correct, the deletion process continues.

The appropriate entry in the entity list is now known. The simplest way to remove it from the list is to change the code number indicating the entity type to one which represents no entity at all. In the examples here we use 0 to indicate a non-defined entity. This process effectively leaves 'holes' in the entity list and it may be necessary, either immediately or at some later stage, to go through some 'house-keeping' process to tidy up the list. If the nodes of the entity are also removed from the node list then holes also appear in this list. If these holes are closed up, then the numbering of the nodes changes and any house-keeping procedure must therefore ensure that the node entries in the entity list are renumbered to correspond with the new node list.

Using the entity list for figure 3.4 (p.52) as an example, we can see that the effect of removing three of the lines and one of the points is to form the modified list given below.

Entity	Code	Node 1	Node 2	Node 3
1	0	1	2	0
2	0	2	3	0
3	2	3	4	0
4	0	4	1	0
5	1	6	0	0
6	3	3	4	0
7	0	7	0	0

In order to remove an entity from the screen when it is deleted, a CAD system will usually redraw it in the same colour as the screen background. This can have the effect of obliterating some part of other entities. Normally, however, some form of 'redraw' command will be provided so that the entire screen picture can be erased and redrawn in its proper form.

Translation and Rotation

Moving on now to manipulations that involve the amendment of the geometric model, we first look at those which move existing entities. These commands are often used to make small changes in a design when entities have been inserted in the wrong place. There are two forms of movement – translation and rotation and, occasionally, one which involves a combination of the two. The important point to note is that movement is an operation which only affects the node list. Once the entities which are to be moved have been identified, the coordinates of the nodes defining those entities are changed in the node list, the entity list remains unaltered.

For example, in order to move a group of entities two units in the x-direction and one unit in the negative y-direction it is only necessary to add 2 to the x-coordinate of each defining node and to subtract 1 from the corresponding y-coordinates. The user therefore needs to specify distances for translatory movements in the x-, y- and z-directions. Often this is accomplished by first identifying the entity that is to be moved, and then using the cursor to indicate two points on the screen (or in space). The system then moves the entity (or entities) parallel to the vector defined along the line joining these two points.

Rotations can also be effected by modifying the node list. To define a rotation, the user needs to specify the centre or axis of rotation and the angle through which the entities are to be rotated.

MOVE AND COPY

Allied to the 'move' commands are the 'move and copy' commands. In order to implement these, the system first duplicates the entity that is to be moved. This is achieved by adding new nodes, with the same coordinates as those involved in the relevant entities, into the node list. These duplicate nodes are then incorporated in the entity list, so creating a second copy of the existing entity. Once the entity has been duplicated, the copy can be moved to its new position by employing the procedures described above. It is, of course, perfectly possible for the system to move and copy whole groups of associated entities in a single operation using this procedure.

'Move and copy' commands are useful for constructing designs which include a lot of similar parts. If, for example, the user is designing a multi-impression die block, it is only necessary to insert the geometry for one impression. After this it can be translated and copied to other places on the block as appropriate. This is one of the ways in which CAD can speed the design/drafting process. It is often no quicker to construct an individual piece of geometry using a CAD system rather than pencil and paper – indeed, it may well be slower. But if that geometry is required in several different locations, then it need only be moved and copied rather than having to be redrawn.

INTERSECTIONS AND TRIMMING

The operations of finding intersections and of trimming are illustrated in figure 3.5. This shows two lines and an arc which are intended to form the bounding edges of a component. At the moment the entities cross each other and, in order to tidy up the design, we need to identify where they intersect and then delete the unwanted parts. Finding intersections is something which is easy to do by eye on a drawing board; on a CAD system, however, it involves the application of numerical algorithms built into the system.

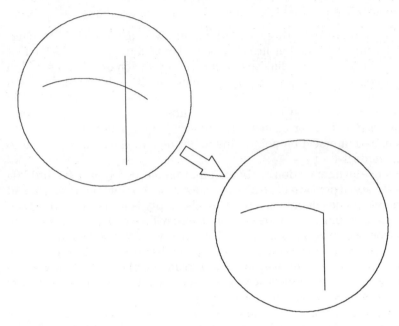

Figure 3.5 Intersections and trimming.

It is not necessary to understand precisely how these work in order to use a system. We can, however, indicate some of the ideas and problems involved by considering the intersection of two straight lines. It is hoped that this will provide some insight into the inherent computational difficulties and may shed some light on why systems can take some time to decide where an intersection occurs and may occasionally produce the wrong answer.

Suppose we have two line segments; one joining nodes with coordinates (x_1, y_1, z_1) and (x_2, y_2, z_2), and the other nodes positioned at (x_3, y_3, z_3) and (x_4, y_4, z_4). If we introduce parameters t and u, then the typical points on each segment are given by the following:

$$(x_1+(x_2-x_1)t, \quad y_1+(y_2-y_1)t, \quad z_1+(z_2-z_1)t\,)$$

$$(x_3+(x_4-x_3)u, \quad y_3+(y_4-y_3)u, \quad z_3+(z_4-z_3)u\,)$$

As t and u vary between 0 and 1, so these points move along the two line segments. If these segments do intersect then at the intersection we have the following matrix equation:

$$
\begin{bmatrix}
(x_2-x_1) & -(x_4-x_3) \\
(y_2-y_1) & -(y_4-y_3) \\
(z_2-z_1) & -(z_4-z_3)
\end{bmatrix}
\begin{bmatrix}
t \\
u
\end{bmatrix}
=
\begin{bmatrix}
(x_3-x_1) \\
(y_3-y_1) \\
(z_3-z_1)
\end{bmatrix}
$$

This represents three equations in two unknowns. It is over-determined. In general, no solution will exist and this reflects the fact that, in general, two line segments will miss each other. If a solution does exist, then it is necessary to check that the point of intersection does in fact lie on both segments. This means that both t and u must lie between 0 and 1. If this is so then substitution of these parameter values into the coordinates of the typical points on both lines gives the required intersection point. This might be added as a new node into the node list.

Once an intersection has been found, it is relatively easy for the CAD software to perform the trimming operation. For lines and arcs, one of the two nodes defining the ends of the entity is replaced by the node corresponding to the intersection. There will, of course, be a choice of one of two end-points and the selection is usually specified by the user in some way. For example, if he is using the cursor to indicate which two entities are to be intersected and trimmed, the parts of the entities that should be removed could be taken to be those nearest to the current cursor position.

STORAGE AND RETRIEVAL

The final part of this module is concerned with two operations performed by a CAD system which do not really fall into any of the three main categories identified above. These are saving geometric information in a disk file and reading it back again. In the simplest form the saving operation involves dumping the entire node and edge lists onto file so that they can be read back directly when required. A more sophisticated approach is to allow the user to structure the data into sub-components or models and to dump each of these to disk separately. This will involve a considerable amount of renumbering of the nodes and entities. Care also needs to be taken when these models are read back to ensure again that further renumbering takes place so that data already in the node and edge lists are not overwritten.

MODULE 3.3 FREE-FORM CURVES 1

For many applications straight lines and circular arcs will be the only geometric entities required. In certain more specialized areas, however, more sophisticated entities will be needed. Tasks such as the design of aircraft wing-sections and of the hulls of ships involve the definition of non-standard surface shapes which are needed to permit efficient operation. Again, the aesthetic appeal of bottles or of plastic packaging containers may depend upon the use of free-form sufaces. In this module we take the first step towards an investigation of such surfaces by looking at free-form curves.

There are a variety of mathematical ways of describing curve forms. What the designer of CAD software is looking for is a means of description that allows a variety of shapes to be generated with computational ease, but which does not require the user of the system to possess an understanding of the mathematics involved.

EXPLICIT AND IMPLICIT FUNCTIONS

One common technique for defining a two-dimensional curve mathematically is to give y explicitly as a function of x. However, even such a common shape as the ellipse is not normally expressed in this way; the usual form of its equation is

$$(x/a)^2 + (y/b)^2 = 1$$

This is an implicit equation relating x and y; we need to rearrange it to obtain y in terms of x. In this case, moreover, we would need to take a square root. We then have problems in evaluating the root and in deciding which sign to take. Additionally, if we are to plot the curve accurately, we would like to be able to take approximately equal steps along its length. But taking equal steps in x does not, for example, guarantee equal steps along the ellipse.

USE OF PARAMETERS

An alternative approach is to introduce a parameter, t, say. The coordinates x, y, z of a point on the curve are written as functions of t and, as t varies, the values of the coordinates change and a curve is traced out. Furthermore, equal steps in the parameters usually produce approximately equal steps along the curve. Because x, y, z are immediately available in terms of t, there are no complicated equations for the CAD system to solve. There is, however one drawback: if we are given a point on the curve it is not obvious which value of t generates it. A related difficulty is that we cannot test easily if any given point in space lies on the curve and finding the intersection of two curves will require special computational techniques.

FERGUSON CUBICS

If we are to write x, y, z as functions of a parameter then we would wish to choose functions as simple as possible. The simplest are polynomials and these are frequently used. The most common degree of polynomial is the cubic, as this is the lowest degree which permits points of inflexion to be generated. Thus, for example, we write x as a cubic in t

$$x = x(t) = a_0 + a_1t + a_2t^2 + a_3t^3$$

Here the coefficients are chosen to produce the desired shape of curve. Usually the parameter t runs between the values 0 and 1. This form of curve is usually attributed to Ferguson (Ferguson, 1964). It has a disadvantage in that it is not clear what the coefficients represent.

BEZIER CUBICS

A variation on the idea has been introduced by Bézier (Bézier, 1972). This amounts to writing the cubic polynomial above in a different way to produce

$$x = x(t) = x_0(1-t)^3 + 3x_1t(1-t)^2 + 3x_2t^2(1-t) + x_3t^3$$

The same procedure is adopted with y and z to give:

$$
\begin{bmatrix} x \\ y \\ z \end{bmatrix} = \begin{bmatrix} x_0 \\ y_0 \\ z_0 \end{bmatrix}(1-t)^3 + 3\begin{bmatrix} x_1 \\ y_1 \\ z_1 \end{bmatrix}t(1-t)^2 + 3\begin{bmatrix} x_2 \\ y_2 \\ z_2 \end{bmatrix}t^2(1-t) + \begin{bmatrix} x_3 \\ y_3 \\ z_3 \end{bmatrix}t^3
$$

or, in terms of position vectors,

$$\mathbf{r} = \mathbf{r}_0(1-t)^3 + 3\mathbf{r}_1t(1-t)^2 + 3\mathbf{r}_2t^2(1-t) + \mathbf{r}_3t^3$$

In these expressions, the parameter t varies between 0 and 1 and, as it does so, a segment of a curve is traced out. What is the advantage of rewriting the cubic polynomials in this Bézier form? At the start of the curve segment we know that t is zero and the above expression gives $\mathbf{r}=\mathbf{r}_0$. Thus $\mathbf{r}_0$ is the start point of the segment. Similarly at the end of the segment, t is unity and the expression gives $\mathbf{r}_3$.

If we differentiate the expression with respect to t, we find that

$$\dot{\mathbf{r}}(t) = 3[\ (\mathbf{r}_1-\mathbf{r}_0)(1-t)^2 + 2(\mathbf{r}_2-\mathbf{r}_1)t(1-t) + (\mathbf{r}_3-\mathbf{r}_2)t^2\ ]$$

This derivative represents a vector along the tangent to the curve at the point where the parameter value is t. Again setting t=0 we find that initially the tangent is parallel to the vector $(\mathbf{r}_1-\mathbf{r}_0)$. As the curve starts at $\mathbf{r}_0$, this means that initially it heads towards the point $\mathbf{r}_1$. In the same

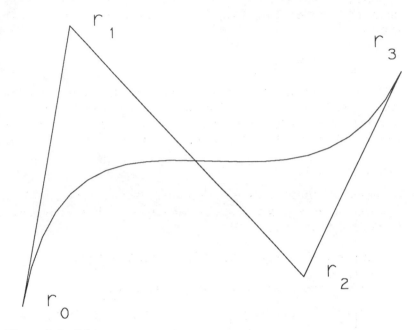

Figure 3.6   Bézier segment and control vectors.

way, setting t=1 shows that the segment ends by arriving at r_3 from the direction of the point r_2. These ideas are summarized in figure 3.6. We have thus found that each of the coefficient vectors in the rearranged Bézier expression has some meaning as far as the shape of the curve is concerned. The coefficient vectors are often referred to as control points or vectors.

The distance of the tangent control vectors from the corresponding end-point determines the extent to which the segment follows the tangent. Figure 3.7 shows the effects of moving one of the tangent control points along the tangent direction. A wide variety of curve shapes can be obtained from the cubic Bézier form; figure 3.8 shows that it is even possible for the curve segment to cross over itself. The control vectors have a further control over the form of the segment; this is the 'convex hull property'. The convex hull of the control vectors is the tetrahedron or (in two dimensions) the quadrilateral that they form. The segment must lie within the convex hull. This is shown in figure 3.9. The significance of this is that if the control points lie in a straight line, then the segment itself lies along that line.

MORE GENERAL BEZIER FORM

Even though Bézier cubics are flexible, there is a limit to the variety of

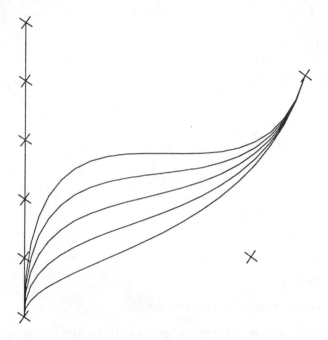

Figure 3.7 The effect of moving one control vector.

shapes that can be produced using them. More flexibility can be
obtained by going to parametric curves of higher degree. The Bézier
segment of degree n has the form

$$\mathbf{r} = \mathbf{r}(t) = \Sigma \begin{pmatrix} n \\ i \end{pmatrix} \mathbf{r}_i \, (1\text{-}t)^{n\text{-}i} \, t^i$$

where

$$\begin{pmatrix} n \\ i \end{pmatrix} = n!/[(n\text{-}i)!i!]$$

is the binomial coefficient and we take 0^0 to be unity wherever it
occurs.

This form has many of the properties of the cubic segment: the
curve starts at $\mathbf{r}_0$ and finishes at $\mathbf{r}_n$; the tangent directions at the ends
of the curve are governed by the first and last pairs of control vectors;
and the curve lies within the region bounded by the control vectors.

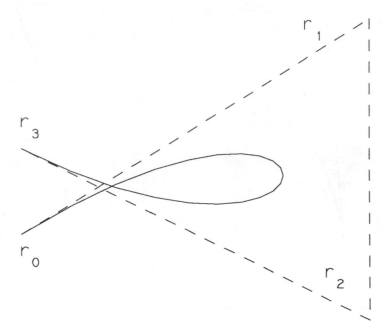

Figure 3.8 A Bézier segment that intersects itself.

PROBLEMS OF HIGH DEGREE CURVES

Moving to higher degree curves can cause problems with oscillations appearing along the segments. A more common approach to gaining a greater variety of curve shapes is to put several cubic segments together. If we are to do this satisfactorily we need to ensure that they join up smoothly. Their end-points must coincide and, by making the last pair of control vectors of one segment collinear with the first pair of the next (in fact two of these points are coincident), we must obtain a smooth change-over of tangent when passing from one segment to the next. This is illustrated in figure 3.10. If we went more deeply into the mathematics we would find that it is also possible to gain continuity of curvature; however this does place further restrictions on the choice of the control vectors.

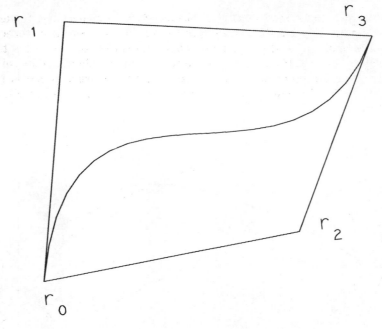

Figure 3.9 The convex hull property.

SUMMARY

In this module we have looked mainly at the Bézier form of smooth curve segment. It allows some geometric meaning to be attached to the vector coefficients used. While many commercial CAD systems use the Bézier form, they very often do not allow immediate access to the control vectors. Instead they permit the user to indicate points on the segment itself and use this information to determine the control vectors and, hence, plot a curve through the given points. Thus an intimate knowledge of how the control vectors work is not necessary. However some understanding of their properties does give the CAD user an insight into what is happening.

MODULE 3.4 FREE-FORM CURVES 2

In this module we look at two variations on the use of parametric curves for defining free-form curves. These are the use of rational curves and of B-spline curves. The former requires the use of homogeneous coordinates and it is this technique that is first discussed.

Homogeneous Coordinates

So far we have restricted the discussion to ordinary Cartesian coordinates – using three numbers to indicate a position in three-dimensional space. Homogeneous coordinates use four numbers to describe a point. If (X, Y, Z, W) are four such coordinates that this represents the point in space whose ordinary Cartesian position is (X/W, Y/W, Z/W). We need, of course, to assume that W is non-zero before

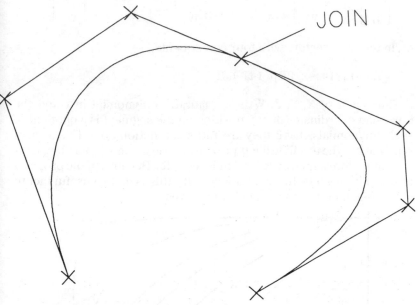

Figure 3.10   Collinear control points to provide tangent continuity.

we can carry out the division. (In fact it is sometimes legitimate to have W as zero, in which case the position represented is a point at infinity.) Note that the homogeneous coordinates for any point are not unique; many sets of coordinates can represent the same point. For example the following all refer to the Cartesian position (2, 3, -1):

$$(2, 3, -1, 1), (4, 6, -2, 2), (-1.0, -1.5, 0.5, -0.5)$$

While this lack of uniqueness may seem confusing, it does give us extra flexibility in obtaining curve shapes. We meet homogeneous coordinates again in the next section when we look at view transformations.

The Bezier Rational Quadratic Form

The next stage is to introduce homogeneous coordinates into the

Bézier form of a curve segment. In order to simplify matters, we reduce the degree of the segments to two for the purposes of discussion. The form of the curve segment is thus:

$$
\begin{bmatrix} X \\ Y \\ Z \\ W \end{bmatrix} = \begin{bmatrix} X_0 \\ Y_0 \\ Z_0 \\ W_0 \end{bmatrix} (1-t)^2 + 2 \begin{bmatrix} X_1 \\ Y_1 \\ Z_1 \\ W_1 \end{bmatrix} t(1-t) + \begin{bmatrix} X_2 \\ Y_2 \\ Z_2 \\ W_2 \end{bmatrix} t^2
$$

or, in terms of vectors with four components,

$$\mathbf{R} = \mathbf{R}_0(1-t)^2 + 2\mathbf{R}_1 t(1-t) + \mathbf{R}_2 t^2$$

Thus each of X, Y, Z, W is a quadratic polynomial in t and the Cartesian coordinates of any position on the segment is a quotient of such polynomials; that is they are 'rational functions' of t. This form of segment has the usual Bézier type of properties. The segment begins at the point represented by $\mathbf{R}_0$ and finishes at $\mathbf{R}_2$. Because of the property of the end-tangents, these intersect at $\mathbf{R}_1$; this is quite a useful fact to note when handling these quadratic curves.

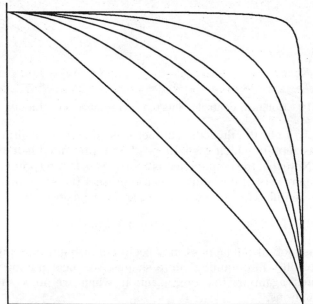

Figure 3.11 Variations in a rational quadratic segment.

Although we are dealing only with polynomials of degree 2 we can gain flexibility by multiplying all the components of any control vector by a non-zero scalar value. This has the effect of changing the curve shape without affecting the Cartesian positions represented by the control vectors. In fact we keep the homogeneous forms for the end-points unchanged and use unity as the fourth component of each. Figure 3.11 shows several curve segments obtained from the control vectors:

$$(1, 0, 0, 1) \qquad (a, a, 0, a) \qquad (0, 1, 0, 1)$$

for various values of a. In particular, when a is equal to $1/\sqrt{2}$ we obtain an exact quadrant of a circle. (More generally, we can generate circular arcs using rational Bézier quadratic segments, taking the control vectors to be the end-points and the intersection of the tangents and choosing the last component of the homogeneous vectors to be unity for the end-points and to be the cosine of one half of the angle between the end radii.) The curves produced by these types of the quadratic Bézier form are precisely segments of conic section curves.

It is possible to work with rational Bézier forms of any degree. However, as the degree increases there is a large amount of flexibility in the choices of control vector positions and of their homogeneous forms. Indeed the amount of flexibility can become confusing and, perhaps for this reason, commercial CAD systems do not usually offer the user these types of forms, explicitly at least.

B-spline Segments

We now turn our attention to B-spline curves. The form of the Bézier segment, as introduced in the last module, can be written as follows.

$$\mathbf{r} = \mathbf{r}(t) = \Sigma\, \mathbf{r}_i\, p_i(t)$$

where the $p_i(t)$ are polynomials and in the case of a cubic segment these are

$$p_0(t) = (1-t)^3 \qquad\qquad p_1(t) = 3t(1-t)^2$$

$$p_2(t) = 3t^2(1-t) \qquad\qquad p_3(t) = t_3$$

It is straightforward to check that the sum of these polynomials is unity, and indeed this holds for the corresponding polynomials for the Bézier segment of degree n. It is this that leads to the convex hull property.

By increasing the degree of the segment, we introduce more control vectors, gaining greater flexibility, but run the risk of introducing unwanted oscillations as a result of the high degrees. Alternatively, we can try to put low degree segments together, but then have to ensure appropriate continuity across the joints. B-spline curves attempt to have a large number of control points to give a lot of control of the curve shape, but do so by keeping the degree low and, in effect, by putting many segments together in a way that preserves adequate continuity throughout.

B-spline Basis Functions

The form of the B-spline segment is as above; its properties depend upon the way in which the function $p_i(t)$ are defined. These are referred to as B-spline basis functions. This is fairly involved and what is given here is only a brief outline. The functions are 'piecewise' polynomials, often of cubic form. This means that over certain ranges of values of t the function is precisely a cubic polynomial; its form changes from one range to the next. The values of t at which the form changes are called 'knots'. It is arranged that the functions are continuous and that their first two derivatives are also continuous across the knots. Additionally the function $p_i(t)$ is made to be identically zero except over four consecutive ranges of values of t.

All these properties are not easy to ensure in general. But we can give a reasonably straightforward example of such a function. We use the idea of Macaulay brackets (often used in the analysis of the bending of beams). These brackets are used to define the following function:

$$[t-a]^n = \begin{cases} 0 & \text{if } t < a \\ (t-a)^n & \text{if} \geq ta \end{cases}$$

Here a is a constant. Thus, for example, when $n=0$ we obtain a unit step function which steps up at $t=a$. When n is greater than zero, this function is continuous and its first $(n-1)$ derivatives are also continuous. Now consider the following sum of Macaulay brackets:

$$([t-1]^3 - 4[t-2]^3 + 6[t-3]^3 - 4[t-4]^3 + [t-5]^3)/6$$

Immediately, from the definition, this function is certainly zero when t is less than one. By multiplying all the brackets out it can be checked that it is also zero for any value of t greater than 5. The use of the Macaulay brackets ensure that it and its first two derivatives are continuous. The knots for this function are 1, 2, 3, 4 and 5. The graph of the function is shown in figure 3.12. The division by 6 in the above formula ensures that the area under the curve is unity.

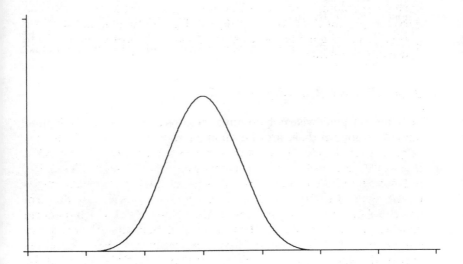

Figure 3.12   A B-spline basis function.

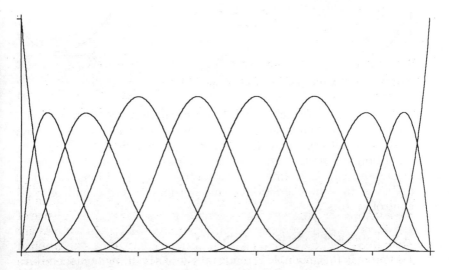

Figure 3.13   B-spline basis functions for the interval $0 < t < 7$.

A<small>N</small> E<small>XAMPLE</small>

When a B-spline curve is defined the set of knots has already been chosen. Usually the first and last knot values are repeated a number of times and these are used to obtain the function $p_i(t)$. If we wish the parameter t to range over values between 0 and 7, then we could take the knots to be:

0 0 0 0 1 2 3 4 5 6 7 7 7 7

and then ten piecewise polynomial functions $p_0(t), ... , p_9(t)$ are defined and the graphs of these are shown in figure 3.13.

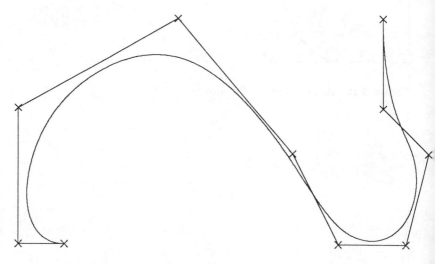

Figure 3.14   A B-spline curve segment.

If ten control vectors are chosen then when these are combined with these functions a curve segment is obtained as t varies between 0 and 7. An example of such a segment is shown in figure 3.14. The repetitions of the first and last knot values ensure that the segment begins and ends at the first and last control vector positions. The initial and final tangents directions are determined by the end pairs of control vectors.

S<small>OME</small> P<small>ROPERTIES AND</small> L<small>OCAL</small> C<small>ONTROL</small>

The curve in this example is made up out of seven segments; one for t between 0 and 1, the next for t between 1 and 2, and so on. Because of the continuity condition on the functions $p_i(t)$ these segments join together smoothly. Since the functions are zero except over four consecutive ranges of values of t, we find that the first segment is influenced only by the first four of the control vectors. The next segment is influenced by the second, third, fourth and fifth control vectors, and so on. In general, for a cubic B-spline, any segment is influenced only by four control vectors and any control vector influences only four segments of the curve. The convex hull property also applies: each segment lies within the convex hull of the four control vectors that influence it. Thus, by taking four consecutive control vectors to be in a straight line, we can produce a straight line portion in an otherwise twisted curve.

Perhaps the most important property of the B-spline segment is that of 'local control'. Because each control vector affects only four segments of the curve, moving it changes only a portion of the full curve. Thus the user of a CAD system can modify a part of an entire curve without disturbing the rest. This property does not apply to Bézier forms even of high order. Figure 3.15 shows the effect of moving one of the control points of a B-spline.

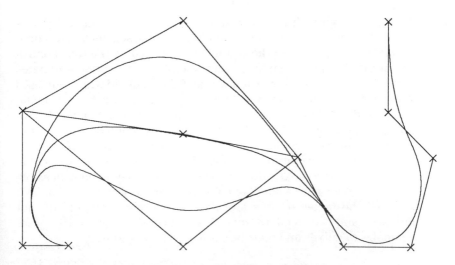

Figure 3.15 Moving a B-spline control point.

MODULE 3.5 FINDING INTERSECTIONS OF FREE-FORM CURVES

In Module 3.2 the intersection of two straight lines was discussed. This is straighforward, as the entities involved are standard and their defining equations are linear. With free-form curves we have deliberately moved away from standard shapes and have introduced polynomials which are inherently non-linear. Thus the problem of creating CAD software to find intersections of such entities is more difficult. To illustrate some of the ideas and difficulties involved, we consider the intersection of a pair of Bézier cubic segments.

ITERATIVE SOLUTION OF SIMULTANEOUS EQUATIONS

The x-, y- and z-coordinates of positions on the two segments are given by cubic polynomials in two different parameters. At an intersection these coordinates are equal in pairs and we could write down three simultaneous non-linear equations connecting the two parameters. This is an over-determined system of equations. No solution may exist and this corresponds to the curve segments not meeting. We could attempt to find solutions by an iterative technique such as Newton-Raphson, but the solution found is highly dependent upon the starting point chosen for the iterations. Difficulties also exist if the segments intersect in more than one place; not all the intersections may be discovered.

BOXES

A different approach is based upon a subdivision technique. Use is made of the convex hull property for Bézier cubic segments. The curve segment must lie within the tetrahedron formed by the four control vectors. If we find the maximum and minimum x-, y- and z-coordinates of the four control vectors, then we can form a cuboid which encloses them. This box has faces which are perpendicular to the coordinate axes and the eight corners have coordinates, each of which is one of the maximum or minimum values. Since the cuboid contains the control vectors it certainly contains the curve segment itself.

Suppose we are trying to find the intersection of two cubic segments. If their boxes do not overlap then they cannot possibly intersect. So we have a quick test for whether the segments miss each other. If the boxes do overlap, then the segments might still not intersect and we need to investigate further. Figure 3.16 shows examples of curves and their boxes in two dimensions.

SUBDIVISION AND THE DE CASTELJAU ALGORITHM

We now come to the subdivision idea. Suppose that the control vectors for a cubic segment are $\mathbf{a}_0$, $\mathbf{a}_1$, $\mathbf{a}_2$, $\mathbf{a}_3$. We form vector $\mathbf{b}_0$ as the average of $\mathbf{a}_0$ and $\mathbf{a}_1$; $\mathbf{b}_1$ as the average of $\mathbf{a}_1$ and $\mathbf{a}_2$; and so on. We next average these new vectors and, continuing in this way, we can produce a triangular array of vectors:

$$
\begin{array}{cccc}
\mathbf{a}_0 & & & \\
 & \mathbf{b}_0 & & \\
\mathbf{a}_1 & & \mathbf{c}_0 & \\
 & \mathbf{b}_1 & & \mathbf{d}_0 \\
\mathbf{a}_2 & & \mathbf{c}_1 & \\
 & \mathbf{b}_2 & & \\
\mathbf{a}_3 & & &
\end{array}
$$

In this array each newly-defined vector is the average of the two to its left. Figure 3.17 shows the positions of these vectors for one example of a cubic segment. It turns out that $\mathbf{d}_0$ is a point on the curve,

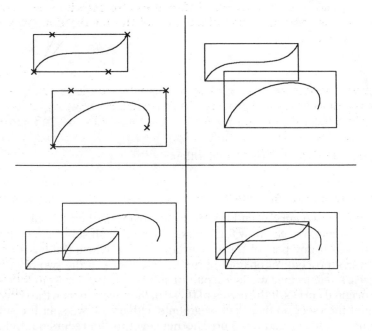

Figure 3.16   Curves and their enveloping boxes.

in fact it is the point corresponding to the parameter being equal to 0.5. This point divides the segment into two sub-segments. Each of these is a Bézier cubic and the control vectors for them are $\mathbf{a}_0$, $\mathbf{b}_0$, $\mathbf{c}_0$, $\mathbf{d}_0$ and $\mathbf{d}_0$, $\mathbf{c}_1$, $\mathbf{b}_2$, $\mathbf{a}_3$; these are precisely the vectors along the sloping sides of the above triangle. This subdivision process is a particular case of the de Casteljau algorithm (see Böhm, 1984).

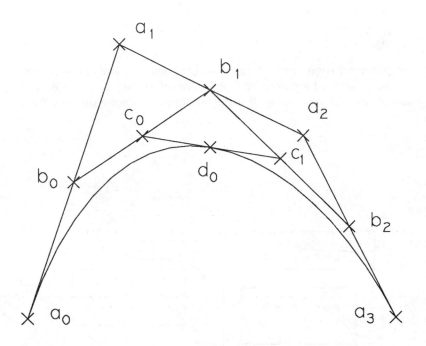

Figure 3.17 The de Casteljau algorithm.

THE INTERSECTION ALGORITHM

If we have two curves, and if their boxes overlap, then we subdivide both of them. We can now form boxes around each sub-segment and try to see if any pair of these overlap. This provides the basis of a recursive procedure for finding the intersections of the two original curves. At each stage we have a pair of sub-segments, one from each of the original curves. If the boxes of these do not overlap then the curves cannot intersect on these sub-segments. Otherwise we split the sub-segments further and descend another level in the recursion. Being recursive, this algorithm is well-suited to implementation in computer programming languages such as Pascal. We can stop the recursion in one of several ways: when we have gone through a prescribed number of levels; when the size of the boxes around the sub-segments is

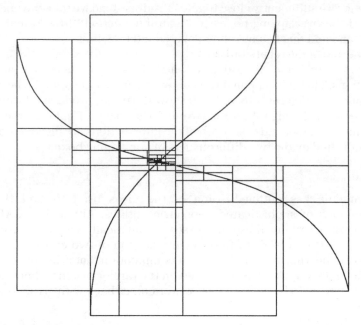

Figure 3.18 Finding an intersection by using the enveloping boxes.

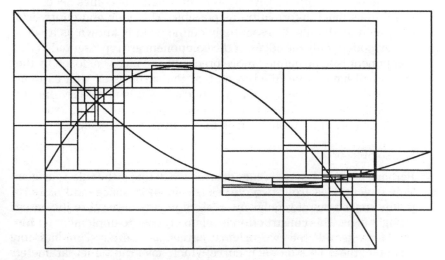

Figure 3.19 Boxes will find both intersections.

smaller than some prescribed amount; or when the parameter values for the original curves corresponding to the ends of the current sub-segments are sufficiently close together. At this stage we take the mid-point of the overlapping boxes to be an intersection of the original curves and add this to a list of discovered intersections.

At the end of the recursion process we may need to examine the list of intersections and discard any which look as though they are repetitions. This can happen if the curves are almost tangential where they intersect. Figures 3.18 and 3.19 show the boxes produced during the course of the algorithm for two different pairs of cubic curves. Note that, in the second case, both of the intersections are discovered, as is indicated by the two different sets of converging boxes.

SUMMARY

This module has outlined some of the ideas used to find the intersections of sophisticated geometric entities. Although a CAD system will carry out these processes automatically when the user indicates that he wishes to find the intersection of two entities, it is well worth bearing in mind the sort of computation that is involved. It can shed light on what has happened on the rare occasions when the system appears to be operating slowly or comes back with erroneous results.

MODULE 3.6 SURFACES

Many design applications require the creation of a detailed description of a component within a CAD system, in which it becomes necessary to fill in as much information as possible. The node and entity lists described in Module 3.1 essentially define what is known as a wire-frame model – only the edges of the component are represented. If the component is to be defined more precisely, the faces enclosed by the edges will have to be filled in so that the surfaces as well as their boundaries are defined. In many applications it is necessary only to deal with planar surfaces, but for more sophisticated work, such as the design of car body shells, free-form surfaces are required.

RULED SURFACES

Perhaps the simplest non-planar surface to define is the ruled surface. This is generated from two curves (or lines) in space, and pairs of points spaced out at regular intervals along the curves are joined by straight lines. This construction is relatively easy to implement within a CAD system. If the two original curves are Bézier segments, for example, the points on each curve which have the same parameter value are joined. An example of this is shown in figure 3.20.

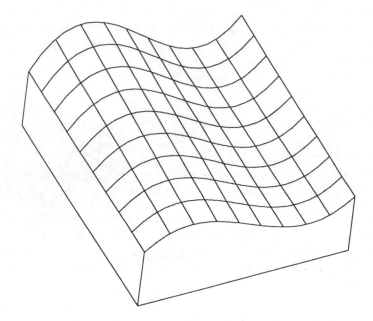

Figure 3.20 A ruled surface.

While it is possible to produce examples in which this process yields unexpected results, it works well in the majority of simple cases. Indeed, the ruled surface is a simple way for the user of a CAD system to build up surfaces. He starts by constructing a curve in a single plane. This is then moved and copied along a line perpendicular to the plane of construction. Fitting a ruled surface between the two curves forms a cylinder (possibly open). An example of this construction process, often called extrusion, is shown in figure 3.21.

BEZIER AND B-SPLINE PATCHES

More sophisticated free-form surfaces can be defined by extending the ideas of Bézier and B-spline curves. We shall concentrate upon Bézier surfaces as these are somewhat easier to appreciate; B-spline surfaces are a similar extension of B-spline curve segments. For curves we defined a Bézier segment – a part of a curve. For surfaces the corresponding concept is the 'patch'. This is also defined parametrically, except that now two parameters are used. The formula for the general point on a Bézier patch is as follows:

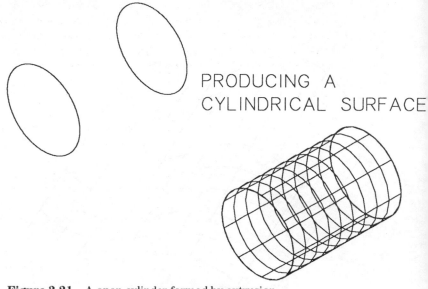

Figure 3.21   A open cylinder formed by extrusion.

$$\mathbf{r}(u,v) = \Sigma\Sigma \begin{pmatrix} n \\ i \end{pmatrix} \begin{pmatrix} m \\ j \end{pmatrix} \mathbf{r}_{ij} (1\mu)^{n-i}u^i(1-v)^{m-j}v^j$$

Here u and v are the two parameters and they both range between 0 and 1. The $\mathbf{r}_{ij}$ are the control vectors for the patch. It can be seen that if v is held fixed and u is allowed to vary, then we produce a Bézier curve of degree n. In the same way if u is fixed and v is varied then a curve of degree m is obtained. In particular, the four boundaries of the patch given when u or v is set to 0 or 1 are each Bézier curves. The curves obtained in this way are called isoparametric curves (since one parameter takes a fixed value) and a number of these are often plotted to indicate the shape of the patch. An example of this is shown in figure 3.22. This has both m and n set equal to 3. Figure 3.23 shows the sixteen control vectors that generate this patch; they form what is called the Bézier polyhedron.

As a particular case of the above general form, consider what happens if m=1. The formula is then linear in the parameter v. In this case the patch represents a ruled surface between the boundary curves corresponding to v=0 and to v=1.

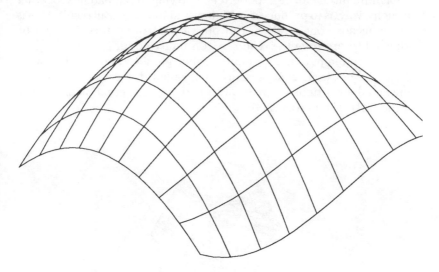

Figure 3.22 A Bézier patch showing isoparametric lines.

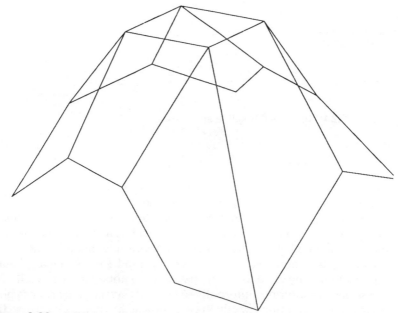

Figure 3.23 A Bézier polyhedron.

PUTTING PATCHES TOGETHER

Great difficulties can be experienced in trying to put patches together in such a way as to produce a sufficiently smooth overall result. This is partly because one tends to run out of control vectors that can be adjusted to preserve the required continuity conditions. Figure 3.24 shows an attempt to use B-spline surface patches to define the outer shell of a bottle. It is an attempt that failed due to the creation of unwanted distortions when the automatic surface-creation procedure of the CAD system was used.

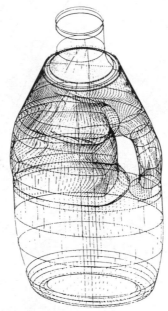

Figure 3.24   Ill-formed B-spline surfaces.

USER INTERACTION FOR SURFACE CREATION

Partly because of this problem, and partly to ease the problem of surface creation, CAD systems provide a number of means of constructing surfaces indirectly, without going to the level of control vectors or even of points on the surface itself. The idea of extruding a curve mentioned above is such a construction technique. Another is the ability to generate a surface of revolution by revolving a (planar) curve about a specified axis. The user is then not concerned with how the surface is described internally to the CAD system. (In order to hold a surface of revolution exactly in Bézier form, it is necessary to use the surface version of the rational curve form discussed in Module 3.4; this is to allow the circular sections to be dealt with.)

Sweeping is a process akin to extrusion. Again a curve is defined, but instead of being moved along a straight line, it is moved along a more general (free-form) curve. Figure 3.25 shows a pipe defined in this fashion. The commercially available DUCT system uses this approach to define volumetric shapes which can subsequently be combined to produce more sophisticated shapes. A basic duct is formed from a closed curve and a 'spine' curve along which the former is swept.

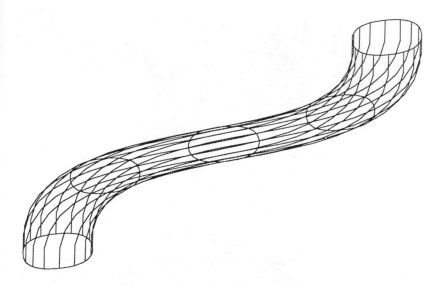

Figure 3.25   Surfaces formed by sweeping.

Figure 3.26 shows a variation on this approach. Here a non-closed curve (shown vertical) is swept around a closed curve (shown horizontal). The curves themselves are composed of rational quadratic Bézier segments. After the initial sweeping stages the various segments involved define the boundaries of faces. For a rational quadratic Bézier patch there are nine control vectors required; eight of these are already known because the boundaries of the faces are available. A special-purpose algorithm is used to obtain a suitable ninth control vector in such a way that, for simple shapes like the one shown in the figure, a sufficient degree of continuity is assured. A particular case of this method is when the 'horizontal' curve is a circle and a surface of revolution is obtained; an example is shown in figure 3.27.

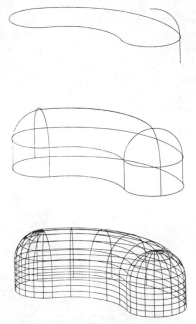

Figure 3.26   Surface patches formed by sweeping.

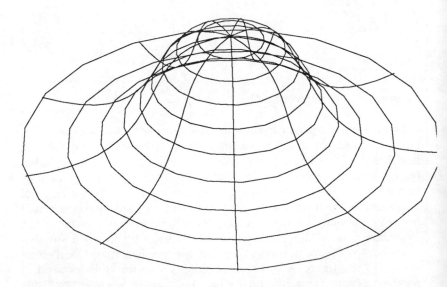

Figure 3.27   A surface of revolution.

EXERCISES

1. Decide what sort of features would be appropriate for a CAD system used for the designing of each of the following types of product. Is three-dimensional modelling required, or will a two-dimensional system suffice? What geometric entities are desirable? What geometric manipulations are likely to be required frequently?

> turbine blades
> office furniture
> multi-impression dies for plastic bottles
> hot forging dies for connecting rods
> cardboard packaging
> printed circuit boards

2. Design and develop a small computer program that will draw two-dimensional shapes made up of straight-line segments. The geometric data should be read into the program from an ordinary text file containing the data in a node and an edge list.

3. The planar curve segment shown in figure 3.28 is an exact Bézier cubic segment. By educated guesswork (or otherwise!) determine its four control points.

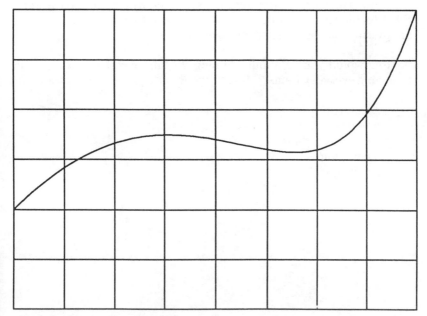

Figure 3.28 A planar Bézier cubic segment.

4. Find an example of a Bézier cubic segment which has a cusp. Can you suggest conditions on the control points for such a segment which would guarantee a cusp and which would guarantee that the segment cuts itself?

5. What are the advantages of box-testing and other recursive techniques for finding the intersection of two curves and/or surfaces? Suggest ways in which the enclosing box round a curve segment might be defined.

6. Design and develop a simple computer program for drawing planar Bézier cubic segments. The positions of the control points could be read from the keyboard or from a text file. The segment can be plotted by evaluating a number of points (say twenty) spaced at equal intervals of the parameter. (How might this procedure be improved upon?) Investigate the effects of changing the various control points upon the shape of a segment.

7. If you have access to a CAD system which supports B-spline curves, investigate the types of shape that can be produced. If it is possible to gain access to the individual control points, look at the property of local control by defining a segment with many control points and then noting the effects produced by causing these to vary.

8. One of the difficulties of finding the intersections of two surfaces patches is that the theoretical intersection curve may not be exactly of the form of any of the curve segments defined within the CAD system. What techniques might be used to overcome this problem and what are the advantages or disadvantages of each?

9. How would you go about defining the surfaces that make up the following everyday items?

> a rolling pin
> a flower pot
> a gingerbread-man cutter
> a car bumper
> a water tap

SECTION 4
VIEW
TRANSFORMATIONS

The conventional engineering drawing comprises a number of two-dimensional representations of a three-dimensional object. These are usually plans and elevations plus, sometimes, auxiliary views of areas of key interest. A CAD system can, of course, be used to produce conventional drawings of this kind, or to display them on the screen; this simply requires that the system should present views of the geometric model along the principal coordinate axes. But, in the case of a system which can hold a three-dimensional model, views can also be constructed that show the model from any other point of view. This can also be done on a drawing board, with the difference that what takes the system a matter of seconds may take the draftsman several hours of painstaking labour. The point is that the arithmetical operations needed to achieve such transformations are straightforward and the computer can process the data quickly enough to provide the user with 'non-standard' views as and when they are required.

When we look at a design along different lines of sight we are effectively using rotations. We can regard the process as being a rotation of the object itself in space, or a rotation of the observer about the object. In the following modules we discuss rotations and a number of other view transformations, including translation and scaling.

Translation, also called 'panning', allows a view (or a model) to be moved about the screen (or in space). Scaling causes the shape of an object to be expanded or contracted in one or more directions. When the same scale factor is used in all directions to a view, we obtain an image which is either enlarged or reduced, this is 'zooming'.

Most CAD systems allow more than one view of a design to be shown at the same time although, obviously, the actual size of each image on the screen must be reduced so that they can all fit on together. This allows the user to have a clearer picture in his own mind of the three-dimensional model held in the 'mind' of the machine. The facility is also useful when it is necessary to point to particular features or entities which may be clearer in one view than in another.

MODULE 4.1 TWO-DIMENSIONAL TRANSFORMATIONS – ZOOM, PAN AND ROTATE

Suppose we are creating a large and complex two-dimensional design on a CAD system. There will come a point at which so much information has to be displayed on the graphics screen that some of it becomes unintelligible: by using the 'zoom up' command we can then, as it were, enlarge that section of the design which we are currently interested in.

Zoom

Zooming up thus allows the user to look in greater detail at some area of the drawing by enlarging it to fill the screen. Inspection or modification of the area of interest can then take place. Naturally, the rest of the drawing is then outside the bounds of the screen and cannot be seen. But if the designer then wishes to see how the additions or modifications he has just made fit into the rest of the design, he can use the 'zoom down' command to reveal more of the model.

Pan and Rotate

The idea of zooming is one of the elementary viewing tranformations that is built into most CAD systems. Another is panning. This allows the user to move the entire drawing relative to the graphics screen. If, for example, the user has zoomed in on a particular part of the design and then wishes to see how that relates to a neighbouring part, then the 'pan' command might be used to move, or translate, the design over so that the next part can be viewed at the same scale.

The third of the two-dimensional viewing transformations that we consider in this module is rotation. This is used less frequently for viewing than the other two, and it simply allows the user to rotate the visible part of the drawing on the screen. Normally this rotation would be about the centre of the screen and through an angle specified by the user – in other words the image on the screen is turned round just as one might turn a plan or drawing round on a desktop so that it can be inspected by someone sitting on the other side of the desk. Some examples of zooming, panning and rotation operations are shown in figure 4.1.

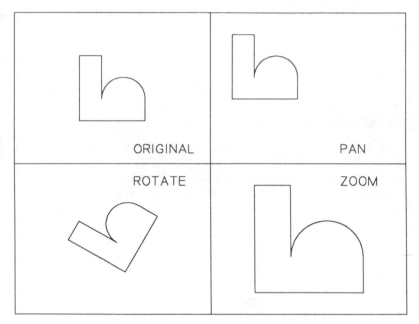

Figure 4.1 Some viewing transforms.

OPERATIONS ON THE NODE LIST

These types of command are simple to use, although the precise way in which they function varies from CAD system to CAD system. The rest of this module looks at how they are implemented in the software.

The viewing transformations operate upon the node list (see Module 3.1), which contains the information defining the geometric positions of the entities. In order to change the view, we need to change the positions of the nodes so that they plot in the desired way on the screen. As we saw in Module 3.1, the node list contains the x-, y- and z-coordinates of each node. However, if these were modified each time a viewing transformation was required, then corruption of the defining data would result. A sequence of zooms up and zooms down, for example, might introduce a series of rounding errors that would significantly reduce the accuracy of the coordinates.

In order to avoid problems of this kind, the node list is enlarged to include new colums containing the plot, or display, coordinates of each node. These can be scaled so that they refer directly to screen positions. Since the screen is a two-dimensional surface, we can assume that there are only two plot coordinates for each node. The original x-, y- and z-coordinates we regard as representing data associated with the 'global space' in which the computer model of the component is defined. These will not often be modified.

In order to plot the entities which make up the screen image we need to apply a transformation that takes the global coordinates of each node and obtains its plot coordinates. This is the viewing transformation. When the view is changed, the transformation is modified, the plot coordinates are recalculated, and a new view of the model is reproduced on a clear screen. Meantime, the defining global coordinates remain undisturbed.

COORDINATES FOR THE DISPLAY

In order to give an idea of the transformation from global to display coordinates, suppose that we are drawing a shape from the node and entity lists and that xmin, xmax, ymin and ymax are the limits on the global x- and y- values of nodes that we wish to appear on the screen. If dxmin, dxmax, dymin and dymax are the screen coordinates which define the bounds of the screen plotting area then we can define scaling factors in the x- and y- directions by

$$\text{scalex} := (\text{dxmax} - \text{dxmin})/(\text{xmax} - \text{xmin})$$
$$\text{scaley} := (\text{dymax} - \text{dymin})/(\text{ymax} - \text{ymin})$$

In order to obtain the same scaling in both directions and so prevent the picture being distorted, we take the actual scale factor to be used to be the smaller of these:

$$\text{scale} := \min(\text{scalex}, \text{scaley})$$

This leaves us with a gap on either the x or the y boundary. This we divide by 2 to obtain a margin on either side and define

$$\text{marginx} := [(\text{dxmax} - \text{dxmin}) - \text{scale}^*(\text{xmax} - \text{xmin})]/2$$
$$\text{marginy} := [(\text{dymax} - \text{dymin}) - \text{scale}^*(\text{ymax} - \text{ymin})]/2$$

At least one of these will be zero if the above definitions are followed precisely. Now the display coordinates for the global two-dimensional position (x, y) are given by:

$$\text{xplot} := (x - \text{xmin})^*\text{scale} + \text{dxmin} + \text{marginx}$$
$$\text{yplot} := (y - \text{ymin})^*\text{scale} + \text{dymin} + \text{marginy}$$

CLIPPING

When these two-dimensional viewing transformations are used, particularly if the user is zooming up on part of the design, some of the nodes will be attached to entities that incorporate other nodes with display coordinates that lie outside the area of the graphics screen. This means that some form of clipping operation has to be performed. So, whenever an entity is to be plotted it is necessary to check if it lies entirely within the borders of the screen, entirely outside them, or only partly within them. In the first case it is drawn, in the second it is ignored, and in the last case some calculation needs to be performed to determine exactly which part should be drawn. Precisely where this clipping operation is carried out often depends on the sophistication of the CAD display equipment. Some screens have built-in software that performs the operation automatically. When this is not the case, the CAD software itself has to do the clipping, utilizing its knowledge of the screen size. In the latter case the time taken to respond to an instruction to perform a viewing transformation will be slightly poorer.

MODULE 4.2 THREE-DIMENSIONAL TRANSFORMATION MATRICES

In Module 3.1, where we introduced the node list, we were careful to provide examples of lists that could store three-dimensional information. The graphics screen, however, can (normally) only display two-dimensional images. In order that the user can see the full three-dimensional effect there has to be some way of allowing him to look at his design from a range of viewpoints. This requires another kind of viewing transformation.

VIEW DIRECTION

The simplest way to indicate to the system which viewpoint should be used is to indicate a 'line of sight', that is to instruct it to 'turn' the model so that one particular aspect is presented to the viewer's eye. In order to achieve such transformations the system must, in effect, perform a two stage operation. First, it takes a copy of the global coordinates of each node and performs a rotation which creates a new set of coordinates such that all the points on the z axis lie along the prescribed line of sight. If the new z-coordinate of each node is then discarded the system is, in effect, left with a set of x and y plot coordinates. The second stage is then to scale these x- and y-coordinates to fit the screen, as described in the previous module.

Once a view direction has been established, the two-dimensional viewing transformations of pan, rotate and zoom can be used as before. Thus the user may first decide to look at his component from a particular view point and then choose to zoom up on it to obtain a better view of some area of interest.

We may additionally wish to use three-dimensional transformations. For example, it may be desirable to rotate the whole model about some slanting axis in space. For the rest of this module we look at a means of describing a range of differing types of transformation in a common way. This approach uses matrices and homogeneous coordinates.

HOMOGENEOUS COORDINATES

A system of homogeneous coordinates, which were introduced in Module 3.4 to allow the definition of rational parametric curve segments, requires four numbers to describe each point in space instead of the usual three. If (X, Y, Z, W) are the homogeneous coordinates of a point, then its Cartesian position is given by (X/W, Y/W, Z/W). As noted in Module 3.4, the representation of any position in this way is not unique.

To allow for the use of homogeneous coordinates, we once more extend the definition of the node list and use four coordinates for the global position of every node. The node list now looks something like this

Node	X	Y	Z	W	X plot	Y plot
1	1.4	2.2	0.5	1.0	856.4	453.8
2	1.1	0.3	2.7	1.0	743.2	634.9
.	.	.	.	.	.	.
.	.	.	.	.	.	.

The simplest way to convert from Cartesian to homogeneous coordinates is to add unity as the fourth coordinate, leaving the other three unchanged. For this reason, the fourth coordinate of many nodes is going to be 1.0. It is in fact entirely possible in many applications to omit the fourth coordinate from the node list and to introduce it only when required. For convenience, however, we assume here that it is always present.

MATRIX TRANSFORMATIONS

We treat the homogeneous coordinates of any node as being the entries in a column vector. Certain of the transformations can then be described by 4x4 matrices which we premultiply onto these column vectors in order to effect the transformation. It is these matrices that we now investigate.

Consider firstly a translation (or pan) through distances a, b and c which are parallel to the x-, y- and z-axes respectively. The effect is to add a, b and c onto the Cartesian coordinates of each node. The following matrix has the same effect for homogeneous coordinates.

$$
T = \begin{bmatrix} 1 & 0 & 0 & a \\ 0 & 1 & 0 & b \\ 0 & 0 & 1 & c \\ 0 & 0 & 0 & 1 \end{bmatrix}
$$

If $[X, Y, Z, W]^T$ is the column vector of homogeneous coordinates of a node (the T superscript here denotes matrix transpose) then the result of premultiplying it by the matrix T is to obtain the column vector

$$[\ X+aW \quad Y+bW \quad Z+cW \quad W\ ]^T$$

and this corresponds to the Cartesian position

$$(\ (X/W)+a,\ (Y/W)+b,\ (Z/W)+c\ )$$

which is the original position translated appropriately. Note that if W is unity then the result is even more obvious.

It is well-known that the following 3x3 matrix represents a rotation through an anticlockwise angle about the z-axis.

$$
\begin{bmatrix} \cos\alpha & -\sin\alpha & 0 \\ \sin\alpha & \cos\alpha & 0 \\ 0 & 0 & 1 \end{bmatrix}
$$

By adding in an extra last row and column we obtain the following 4x4 matrix

$$
R = \begin{bmatrix} \cos\alpha & -\sin\alpha & 0 & 0 \\ \sin\alpha & \cos\alpha & 0 & 0 \\ 0 & 0 & 1 & 0 \\ 0 & 0 & 0 & 1 \end{bmatrix}
$$

It is straightforward to check that the matrix R also represents the same rotation when the positions concerned are represented by homogeneous coordinates. Indeed if P is any 3x3 rotation matrix for Cartesian coordinates then the following is the corresponding 4x4 matrix for use with homogeneous coordinates

$$
\begin{bmatrix} & & & 0 \\ & P & & 0 \\ & & & 0 \\ 0 & 0 & 0 & 1 \end{bmatrix}
$$

In order to bring the user's line of sight along the z-axis we use a 4x4 rotation matrix. If u, v and w are the Cartesian components of a unit vector along the line of sight, and we set

$$d = (1-w^2)^{1/2} = (u^2 + v^2)^{1/2}$$

and assume that d is non-zero, then the required rotation matrix is:

$$\begin{bmatrix} -v/d & u/d & 0 & 0 \\ -uw/d & -vw/d & d & 0 \\ u & v & w & 0 \\ 0 & 0 & 0 & 1 \end{bmatrix}$$

(This leaves the original z-axis in the new yz-plane so that when the new z-coordinate is discarded for plotting purposes the original z-axis looks vertical on the screen. The matrix needs to be chosen specially if d is zero; this corresponds to viewing along the original z-axis.)

Finally the following 4x4 matrix permits scaling in each of the Cartesian directions by amounts s_x, s_y, s_z.

$$\begin{bmatrix} s_x & 0 & 0 & 0 \\ 0 & s_y & 0 & 0 \\ 0 & 0 & s_z & 0 \\ 0 & 0 & 0 & 1 \end{bmatrix}$$

By taking the scale factors all to be the same, this matrix represents a zoom transformation.

A SINGLE FORM OF TRANSFORM

The advantage of using homogeneous coordinates is that all these transformations can be represented in a common way – by using 4x4 matrices. Thus transformations can be combined by multiplying their matrices together. When using these matrices for viewing transformations, the sequence of operations is as follows. Firstly, we transform the global coordinates of a node by applying the transformation matrix. We then discard the z-coordinate, evaluate the Cartesian x- and y-values (by division by the last homogeneous coordinate) and convert these to display coordinates as in the previous module. If the associated entity survives the clipping process, these values are then used for drawing on the screen.

MODULE 4.3 AXIAL AND OBSERVER SYSTEMS

In the previous module we looked at transformations and, in particular, at how a rotation matrix could be used to bring a specified axis through a model along the user's line of sight and thus provide a different view of a three-dimensional object.

Line of Sight

The use of the line-of-sight direction in this way is one means of defining a view. Indeed, the views along the three coordinates axes, which provide the usual plans and elevations of an engineering drawing, can be thought of simply as three commonly specified lines of sight. A fourth is the view along the line joining the point $(1,1,1)$ to the origin of the coordinate system which creates an isometric type of view. Many CAD systems have these views built into them so that they can be called up directly when required.

In general, however, in order to specify the use of a particular line of sight, the user must be able to specify points along the desired axis. This may require him to have knowledge of the position of the origin and the orientation of the coordinate system being used or to have the ability to specify viewing points in space not necessarily related to the design itself. This may be perceived as a drawback and CAD systems very often let the user specify non-standard views in a different way, in terms of rotations of the view from its current or its absolute global position.

Viewing Rotations

The simplest way to achieve this is to provide for viewing rotations about each of the main coordinate axes. As the last module showed, it is a straightforward task for the system to construct the corresponding rotation matrices. But while it is possible to generate any rotation about any axis as a combination of rotations about the main axes, the process can be very confusing for the user. One problem is that sequences of rotations are not commutative; that is to say, the order in which the rotations are performed is significant – a rotation about the x-axis followed by one about the y-axis does not, in general, produce the same result as the same rotations performed in the reverse order. Another difficulty is that it is easy to forget the orientations of the axes after a few rotations have been performed and so request the system to perform an unwanted transformation. The user does not discover his error until the view has changed and may by then have forgotten what transformation he tried.

Observer Coordinates

An alternative to using rotations about the axes is to use an 'observer' coordinate system. This is illustrated in figure 4.2. In order to transform the current view to the desired one the user specifies three rotations. The first is about the vertical axis of the existing view; this is followed by a rotation about a horizontal axis that brings the 'eye' up to the appropriate viewing position. This defines a line of sight. Finally there is a rotation about the line of sight itself which is equivalent to a rotation of the two-dimensional view as discussed in Module 4.1.

The rotation transformation matrix that effects these three separate rotations in the order specified and based on the axis system shown in figure 4.2 is as follows:

$$\begin{bmatrix} \cos\cos + \sin\sin\sin & -\sin\cos & \cos\sin + \sin\sin\cos & 0 \\ \sin\cos - \cos\sin\sin & \cos\cos & \sin\sin - \cos\sin\cos & 0 \\ -\cos\sin & \sin & \cos\cos & 0 \\ 0 & 0 & 0 & 1 \end{bmatrix}$$

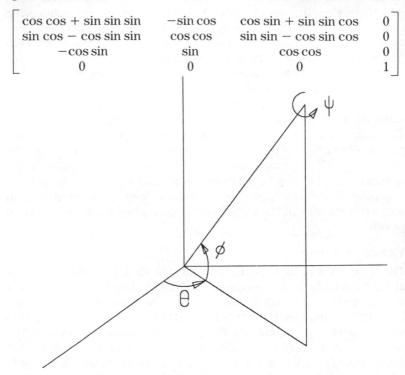

Figure 4.2 Viewer Coordinate System

COMPOUND TRANSFORMATIONS

When the user runs through a whole sequence of view transformations in order to obtain a particular view, it is, of course, necessary for the CAD system to keep track of what is happening. One way for it to do this is to create and update a view transformation. It is this that is used in the first stage of the transformation of the global coordinates of the nodes into display coordinates. (The other stage is the screen transformation which adjusts the two-dimensional coordinates to fit the graphics device.) If a sequence of view transformations is carried out, then the system simply premultiplies the current view transformation at each step by the matrix for the new transformation. If the user wishes at any stage to go back to one of the standard views (or to a previously defined view), then the current view transformation is simply overwritten by the new one.

The order in which these viewing and display transformations are applied is shown, in summary, in figure 4.3. Here the clipping of the image to fit the graphics screen (the procedure described in the last module) is shown as the final stage in the operation. This does, however, mean that if much of the current plot is to be left out of the next view (as is, for example, the case when an image is zoomed up), a great deal of data will have been transformed and then not used. Time can therefore be saved by first finding the volume (in the global space of the model) which transforms on to the rectangle which is the border of the final view. The geometry is then clipped against this volume and only that which is to be plotted actually has the view transforms applied to it. In this case, the clipping operation would precede the others, see figure 4.6.

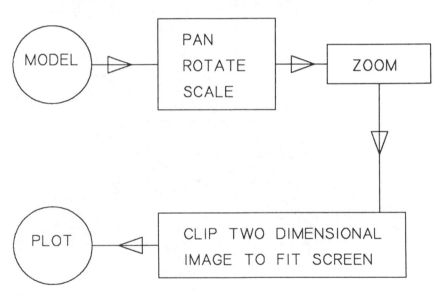

Figure 4.3   Order of viewing transformation.

MODULE 4.4 THE USE OF PERSPECTIVE

The last two modules have looked at ways of transforming the view of the design on the screen so that its full three-dimensional nature is revealed. Another way of enhancing the three-dimensional effect is through the use of perspective. This, too, can be achieved by the use of a 4x4 transformation matrix and it therefore fits in neatly with the types of viewing transformation already discussed.

THE DEPTH COORDINATE

The view transformation techniques described in previous modules have all involved, as part of a first stage, the application of a rotation that brings the line of sight of the user along the z-axis. The point being that once this has been done the z-coordinate can, in effect, be discarded, leaving just the x- and y-values to be plotted, after they have been scaled so as to fit on the screen. But a perspective view will, of course, make use of the z-value as well. There are several different forms of perspective that can be used within CAD systems. The one described here is relatively simple, but it provides a perfectly adequate three-dimensional illusion.

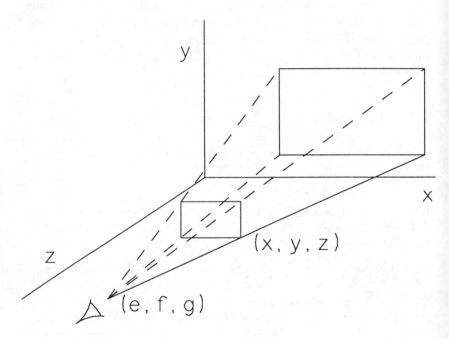

Figure 4.4   Definition of perspective.

THE INTRODUCTION OF PERSPECTIVE

In order to obtain a perspective view from any particular angle, we start by assuming that the model has already been rotated so that the eye is looking along a line parallel to the z-axis. This situation is shown in figure 4.4. The eye is at the position with coordinates (e, f, g). We now consider what the eye 'sees' by projecting the nodes onto the xy-plane (where $z=0$). A typical point (x, y, z) is shown in figure 4.4. We construct a straight line from the eye, through the point, to meet the xy-plane. A little coordinate geometry shows that the point where it meets the plane has coordinates given by

x-coordinate = $(x-ehz)/(1-hz)$
y-coordinate = $(y-fhz)/(1-hz)$

where $h= 1/g$ and assuming, of course, that g is non-zero so that the eye is not on the plane.

MATRIX FORM

Thus we wish to transform each set of coordinates (x, y, z) in the current view to

$((x-e)/(1-hz), (y-f)/(1-hz))$

before we use the screen transformation and plot the results. This perspective tranform can be described by a 4x4 matrix and we can use this provided we retain the homogeneous coordinates. The appropriate matrix is the following.

$$\begin{bmatrix} 1 & 0 & -eh & 0 \\ 0 & 1 & -fh & 0 \\ 0 & 0 & 1 & 0 \\ 0 & 0 & -h & 1 \end{bmatrix}$$

Premultiplication of this onto the column vector $[X, Y, Z, W]^T$ gives the column vector

$[X-ehZ \quad Y-fhZ \quad Z \quad W-hZ]^T$

When we divide the first two components by the last we obtain the following values which are precisely the required expressions for the coordinates after the application of perspective

$[(X/W)-eh(Z/W)]/[1-h(Z/W)]$
$[(Y/W)-fh(Z/W)]/[1-h(Z/W)]$

Figure 4.5 shows views of the same object with varying values of h so that effect of different distances of the eye from the plane can be seen.

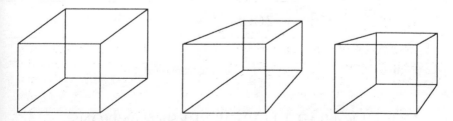

Figure 4.5 The effect of different viewing distances on the images produced.

We note that when h is zero, so that g is infinite and the eye is very distant from the plane, then the above matrix represents an ordinary translation transformation which carries the point (e, f, 0) at the end of the line of sight to the origin.

EFFECTS OF THE EYE COORDINATES

It might be thought that the only important piece of information about the eye position is its distance, g, from the plane. However e and f do also affect the form of the view obtained. It should normally be the case that the image of (e, f, 0), the point at the end of the line of sight, is actually within the borders of the graphics screen. If this is not so, then the lines indicating the perspective converge to a point off the screen and the result is confusing to the user. Thus it is important to know precisely which line of sight is being used and not just its direction.

As already indicated, we need to apply the perspective transformation after the rotation to achieve the desired line of sight and before the screen transformation. Thus figure 4.3 which was shown at the end of the last module is now extended and replaced by figure 4.6.

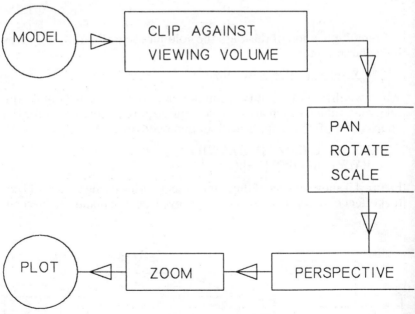

Figure 4.6   Revised order of the viewing transformation.

MODULE 4.5 MULTI-VIEW PRESENTATIONS

A conventional engineering drawing is composed of a number of views, usually including a plan and elevations. At least two views are required to convey a reasonably complete description of any object. These are simpler to construct than oblique angle views; but, even so, considerable skill is required both to read such a drawing and to construct it in the first place.

Two-dimensional CAD

A two-dimensional computer-aided drafting system can be used to create a range of different views of the same component. But the task is carried out in essentially the same way as it would be on a conventional drawing board. Each view is composed of lines and other geometric entities and no view has any relation to any other. If a line is removed from one view the corresponding entity will still remain in place in all the other views. Thus, although it is created by a computer, the final drawing, when plotted out with its different plans and elevations, is no more likely to be correct and consistent than if it had been done by purely manual means.

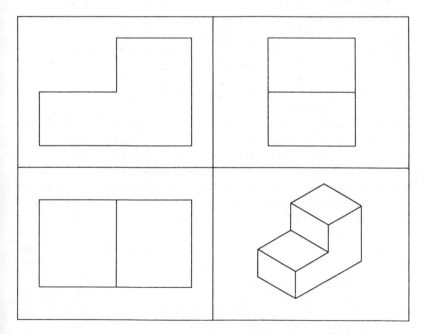

Figure 4.7 Three orthographic and one isometric view of a block.

Three-dimensional CAD – only one model

With a three-dimensional CAD system, however, we are in a somewhat different position. Now just a single model of the component being designed is created. This can be viewed from different points in space, but each view relates back to the single model. So if we remove a line from one view, it will be removed from the model and will no longer appear in any other views. This means that there is far more likelihood that all views of a component will be consistent with each other.

MULTIPLE VIEWS FROM THE MODEL

In order to give the user a feel for the three-dimensional nature of the design, most CAD systems allow more than one view to be displayed at a time on the graphics screen. To do this the screen is divided into a number of areas called 'windows' or 'viewports' and the user selects a different view to appear in each. These may be views along the main axes (so that the standard plans and elevations are obtained), isometric views or views along oblique lines of sight. It may also be possible to zoom up on one or more of these views so that critical areas are highlighted in one window while the overall view is seen in another. Naturally, the images themselves will be smaller as there is less screen space in which to display them. Figure 4.7 shows several views of the same object displayed in this sort of way.

It is normally possibly to lay out the various views on the screen and then to output them to a plotter. In this way a more or less conventional engineering drawing can be created, should this be necessary, and, since the views are generated from a single model, each will be consistent with the others.

CAD systems vary in the way in which they handle dimensions and other textual information. Some, in effect, incorporate it into the model of the component itself and, as a result, it appears in all views – sometimes in very strange orientations. A better approach is for the system to regard this type of data as belonging just to the view in which they have been created and to display them only in that view. Should the user wish the data to appear elsewhere he needs to insert them as and when required.

In order that the system should be able to generate several views simultaneously, it is necessary to extend the node list yet again, since it needs to have sufficient extra columns to hold the screen display coordinates for each of the views concerned. Additionally the view transformations for each need to be held so that the global coordinates can be converted to generate each plot.

USER INTERACTION WITH MULTIPLE VIEWS

When several views are displayed, some systems expect the user to nominate one of them as the view which he wishes to work on. If he then wants to enter new entities or modify existing ones he must do so on this view. If the user wants to identify an existing entity by using the cursor, then the system will of course determine the cursor position in terms of screen coordinates and these will then be compared with the nodal display coordinates for the view in question. In this way the nearest entity can be identified, working solely in terms of display information.

If, however, the user is trying to indicate a new point in space, then it is necessary to take the screen position of the cursor and apply the inverse of the screen and viewing transformations to obtain a point in the global space of the model. In order to do this the system will have to be provided with a value for the third coordinate. The simplest method is to have the user input a value which represents the distance of the point in front of, or behind, the view on which it has been selected.

More advanced systems allow the user to work in any of the views displayed without identifying any one in particular. This means, for instance, that it is possible to insert a line by indicating one end-point in one view, perhaps an enlarged part of a critical area, and the other end-point in a different view. Since the entity goes into the single model of the component, all views are then updated so that the new line appears in all of them, or at least those in which it is visible.

MODULE 4.6 ADVANCED VIEWING TECHNIQUES

The viewing techniques described in the previous modules are about the best that can be achieved using what is (at the time of writing) the basic type of graphics screen display. They allow a design to be displayed in one or more views and, in some cases, also allow perspective to be added. In general, this will be quite sufficient to convey the illusion of a three-dimensional object.

More advanced screen hardware does, however, make it possible to go even further and some of the ideas employed by these devices are discussed in this module. As time goes by, those features which prove to be most successful will doubtless becomes standard on the graphics terminals of commercial CAD systems.

INTENSITY CUEING

The first of these advanced techniques is intensity cueing. This requires a graphics screen which is capable of displaying points and entities in a varying range of light intensity; in the case of a monochrome screen, this is often referred to as varying the 'grey level'. Such screens are in fact relatively inexpensive and it is perhaps surprising that not many systems make use of them. The basic idea is to show those parts of a design which are 'further away' from the eye at a lower intensity than those which are 'nearer'. This gives the effect of the component disappearing into the distance and its three-dimensional nature is strikingly suggested.

In some of the earlier modules the z-coordinates have been discarded once the viewing transformations have been applied to bring the line of sight along the z-axis pointing out of the screen. In systems offering intensity cueing, the z-coordinates are used to determine the distance of each node behind the screen, and hence the level of intensity to be associated with it. Some computation needs to be carried out since, in general, all parts of an entity will not be equidistant from the eye. These calculations can be based on information about the positions of the end-nodes and on knowledge of the type of entity being considered. Such systems will use several levels of intensity to display a single piece of geometry and may also allow the user to alter the intensities associated with entities at different distances. This can be useful in that it allows the effect to be exaggerated if necessary, or to be eliminated (almost) entirely should a more normal type of display be required.

GRAPHICS PROCESSING IN THE WORKSTATION

With the advent of the workstation, the individual CAD user can now have substantial computing power available at his own terminal. This means that complex computations can be carried out locally without the need to have recourse to the central computer running the CAD software, and so without affecting other users of the system. In many cases the workstation's built-in computational resources will be used to perform the viewing transformations. This means that the appropriate parts of the node and entity lists are effectively sent straight to the workstation and it deals with the viewing transforms and with the subsequent display. In this way the computing load is distributed among many processors.

Because the load on the workstation processor comes only from a single user, it is possible to carry out sequences of transformations very rapidly. Advanced workstations permit this to be done in real time so that a continuously moving (translating and/or rotating) image is seen. Perspective can be added if required. The movement of the display is controlled by the user, either with the help of the tablet and keyboard or via a special set of display control knobs which govern the amounts of rotation and translation about and parallel to various axes.

REAL-TIME TRANSFORMATIONS

At the time of writing, a workstation which can genuinely perform the viewing transformation in real time needs to be very powerful; indeed, it requires a dedicated processor on the scale of a mini-computer. Less powerful workstations go for a compromise solution. The user enters a special viewing option. When this happens, the station processes the node and entity information that it holds and produces transforms of these for a very large range of possible views. Naturally, this preparation takes a perceptible time; but once it has been completed realistic moving displays can be generated by running through the appropriate range of transforms.

STEREOSCOPIC TECHNIQUES

The use of red and green anaglyphs is a long-established method of producing three-dimensional effects. An object is drawn twice, once in red and once in green, as it would appear when seen from two slightly different directions. The user wears a special pair of spectacles which have the eye-pieces tinted red and green. Thus the eye with the red filter sees only the green lines and vice-versa, and this means that each eye can be given the view that it would receive if the three-dimensional object were real. This principle can also be employed in conjunction with a display screen. Two different viewing transformations are used and the resulting views are displayed one on top of the other in the two colours.

This approach has the disadvantage that the image perceived by the user will be a monochrome one. A more recent variation makes use of another kind of special-purpose spectacles with eye-pieces composed of liquid crystals which can be made to go clear or opaque in very rapid succession. Using the speed of a CAD workstation, two views of the same design can be shown in equally rapid succession, one for each eye. The eye-pieces are arranged to 'shutter' on and off alternately so that each eye is only allowed to see its own image of the component. At present this approach is still very much in its infancy. Some interest has been expressed in using it for applications in plant layout, where the problem of ensuring that the various parts do not overlap or clash is acute.

GENUINE THREE DIMENSIONS

Finally we should note that there have been one or two attempts to build display devices which show an object in three dimensions without need for the user to wear special spectacles or the like. Some of these have been weird and wonderful! One of the more successful (and expensive) uses a varifocal mirror. This is a concave, silvered membrane whose focal length can be altered very quickly by means of an arrangement like a conventional loudspeaker. The mirror reflects an image generated by a cathode ray tube, and this produces a virtual image at a distance from the user, a distance which changes as the focal length of the mirror alters. If the cathode ray tube generates, in very rapid succession, a sequence of images each of which shows only those parts of a model which are equidistant from the eye, and if the focal length of the mirror is changed for each image, then a complete virtual image of the object is built up in three dimensions.

Although highly ingenious, such techniques seem unlikely to come into common use. There are already various ways in which a CAD system can suggest the three-dimensional nature of the model it contains and, as far as the user is concerned, a completely accurate solid view is not actually needed. All he requires is something which will allow him to continue his design with confidence. Moreover, the need to wear special apparatus or sit in a particular position may easily become irksome. Realistically, the increasing use of powerful workstations, with their ability to perform real-time changes in viewing displays, seems to be the most promising way ahead for future CAD systems.

EXERCISES

1. Why is it important to apply viewing transformation to a copy of the original geometry held in the database rather than to operate on the original information?

2. Design and devlop a simple computer program to illustrate the use of viewing transformations. Hold the geometry in node and edge lists and add extra 'columns' to the node list so that the view coordinates can be held together with the original, undisturbed model coordinates. Read the geometric data in from a file. Initially read in the transformation matrix from a file or from the keyboard, apply it to the entries in the node list and then plot the view. (You may need to apply clipping to the two-dimensional view if this is not carried out by default by your computer or screen hardware.) Extend your program so that it computes its own transformation matrix from a user-defined line of sight.

3. We are mainly interested in applying view transforms to the geometry of a design. However a design consists of more than just geometry; it includes also dimensions, other textual information, symbols and cross-hatching. Often these items are held in the database along with the geometric entities that make up the design proper. What problems then confront the designer of the CAD system itself? How would you like to see text handled when a view is zoomed up or down, or when it is rotated (and again what problems exist)? How is it actually handled on CAD systems to which you have access?

4. Often a model is clipped against a 'viewing volume' in space. This volume is arranged to transform onto the plotting area of the screen if

the view transformations were applied to it. Why is this procedure more efficient than clipping the two-dimensional image against the screen boundaries? Normally the viewing volume is an infinite rectangular tube in space; what shape would it have if perspective were used? What effect would be produced by closing the ends of the tube to produce a volume of finite size and what advantages would this have?

5. Some modern graphics screens perform viewing transforms themselves. The host computer passes the full three-dimensional coordinates directly to the graphics hardware being used. What advantages does this form of distributed computing have? What are the possibilities for real-time rotations (and other transformations) and/or animation?

SECTION 5
TYPES OF CAD MODELLING SYSTEMS

A drawing board is used by a designer to represent in two dimensions a component that will ultimately exist in three. The ways in which this representation is constructed and presented have developed over time. A number of techniques have evolved and a variety of drafting conventions and standards now exist. The first of the modules in this section discusses some of these approaches.

With a good CAD system, a three-dimensional model of the proposed component is built up within the computer system, although, to display it, we usually have to resort to a two-dimensional medium in the shape of either the graphics screen or the plotter. We can, however, make use of the speed of the computer's arithmetic to create a whole range of additional representations of the design that would not normally form part of a conventional engineering drawing. CAD thus makes it far easier for the designer to convey three-dimensional information about a component.

But the real advantages of using a CAD system does not derive from its capacity for producing sophisticated pictures of a design, especially as it actually takes more time to create a drawing from scratch using CAD than it does by hand. The time benefits of CAD only become evident when we consider how the design information is used further 'downstream', as the basis for analysis, manufacturing and inspection activities. Ultimately, it is the needs of these activities which dictate the form in which the design information should be held. This form also influences, and is influenced by, the ways in which we want to represent the design information graphically. In the last three modules of this section we shall therefore be discussing the various ways in which geometric information can be stored and displayed by a CAD system.

At the simplest, and cheapest, end of the spectrum comes the wire-frame model. This comprises the nodes and edges of the component. It is not a perfect model of the full geometry of that component, but may well provide all the information needed for design and for the subsequent downstream processes. Next comes surface modelling. Here extra information about the bounding faces of the component is held and the system therefore knows more about its geometry. Finally, there are solid modellers which aim to provide a *complete* description of the geometry of the component. They know precisely which parts of space are occupied by the component - that is, which parts of the model represent "solid" and which represent the surrounding space.

Other writers have identified the three main classes of three-dimensional modelling systems mentioned in the last paragraph and have also distinguished between various sub-classes. The definitions used have passed into the jargon of CAD and we shall therefore, describe the classes under the same headings. However, it is important to bear in mind that the different approaches to modelling are not mutually exclusive. Any particular system may exhibit aspects of several different classes. As systems continue to develop other approaches will doubtless be introduced and the distinctions between existing procedures will become more blurred.

MODULE 5.1 TWO-DIMENSIONAL DRAFTING PRACTICE

The purpose of an engineering drawing is to provide an unambiguous description of an item in order that it can be manufactured. The information required is conveyed in a combination of graphical, symbolic and textual forms. Although most of a designer's training is normally concentrated on the development of the skills necessary for the construction of the pictorial views, this is only part of the complete communications activity. Care must be taken, whilst conventions and practices are observed, to ensure that adequate and correct instructions are provided for manufacturing purposes. These may include descriptions of materials, processes, finishes and the relationships between the various components that make up an assembly.

ORTHOGRAPHIC PROJECTION

In order to represent a three-dimensional object upon a two-dimensional plane, it is necessary to construct a number of views from differing positions. Dimensions, in full three-dimensional space, can then be inferred from a combination of views.

The first, and most general, conventions to be applied in drafting are the rules of orthogonality. Here perspective is eliminated. For, while we naturally observe and record objects as appearing to diminish in size as they recede from us, this does mean that the actual dimensions of an object can only be determined from a drawing if the relationship of the drawing plane to the object is known. Move the plane towards or away from the object and a new image will be formed. Such pictorial representations give a very realistic view but are difficult both to construct accurately and to interpret dimensionally.

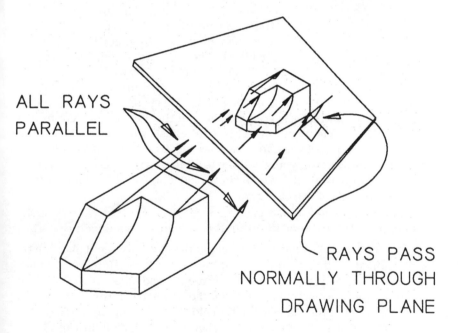

ALL RAYS PARALLEL

RAYS PASS NORMALLY THROUGH DRAWING PLANE

Figure 5.1 Orthographic projection.

The use of an orthographic projection overcomes these problems. This requires that a ray cast from any point on the object will always pass normally through the drawing plane. Such a requirement imposes a number of restrictions upon the drawing construction (figure 5.1). Firstly, for a chosen view plane, there can be only one projection point on that plane for each point on the object. Secondly, the rays cast from all points are parallel. Finally the size of the drawn image do not change as the plane is moved towards or away from the object (as long as the orientation remains the same).

ORTHOGONAL PLANES

To provide sufficient information to allow the size of a component to be calculated in three-dimensional space, it is only necessary to provide two orthographic views in a known relationship to each other. This does, however, create difficulties when dealing with complex objects and can lead to errors. It is therefore more usual to use an orthogonal set of axes and provide a number of views in the reference planes. Thus we are able to create the conventional front view, side view and plan. These are not only at right angles each to the others, but are also set in the planes of the reference dimensions.

The arrangement of these views upon the paper (or plane) is governed by a further set of conventions. Although information could be gathered from each of the views independently, no matter how they were laid out, it is far more convenient if corresponding features align, either horizontally or vertically, between views. Details can then be traced, by set square and T-square, from one view to the other.

It is thus conventional to select a viewing position that can naturally be described as the 'front' of the object. This is usually considered to be the view in which the greatest detail of the object is evident from a 'natural' viewing position. This can cause confusion. From the draftman's point of view, for example, the 'front' view of a car is not a drawing of what we would call the front in everyday terms, but rather the view from the 'side', which shows the length of the vehicle normal to its direction of travel.

FIRST AND THIRD ANGLE PROJECTIONS

Once the front view has been selected the side views and plan can be positioned so that features align between views. Here yet another convention, or practice, has to be recognized. Just as the rules of orthographic projection ensure that the geometric form of the view is independent of the plane's position along its viewing ray, so they also dictate that it is independent of whether the plane is in front of or behind the object. Thus, if an observer looks down the viewing ray towards the object, the view is the same no matter whether it is thought to have been drawn by the observer onto a transparent screen set in front of the component or by the observer using the object to cast shadows onto a screen set behind it. The side view can thus be set either to the side from which it is observed or to the other. When it is drawn to the far side (by casting shadows) it is known as 'first angle projection'. This is because the object can be considered as sitting in the 'first sector' between the front and plan planes upon which the views are to be cast (figure 5.2), that is above and to the right of the axis. Similarly when the view is set to the same side as the observer, the object can be considered to be below and to the left of these construction planes so that the layout is referred to as 'third angle projection' (figure 5.3).

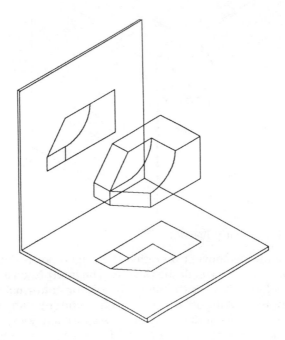

Figure 5.2   First angle projection.

Either construction technique can be employed but they cannot be mixed. Usually, however, it is more convenient for a company to standardize on one system and to incorporate it into its procedures. Indeed, various industries have settled upon one or the other as part of their codes of practice. The national standards which exist in all industrial countries may also include recommendations on which convention to adopt. The British Standard on 'Engineering Drawing Practice' (BS308) is however not prescriptive, allowing either to be employed.

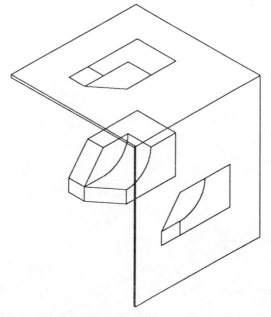

Figure 5.3 Third angle projection.

The choice of one convention rather than the other is determined largely by the ease with which the views can be interpreted as a three-dimensional object. It seems natural to look for confirmation in an auxiliary view by 'looking on' in the same direction (first angle) rather than 'looking back' (as in third angle). This approach does, however, start to breakdown once the view one is looking across becomes large, since it is not easy to compare remote geometric features with any degree of certainty. Third angle construction usually allows selected elements in differing views to be arranged at their closest. It is for this reason that third angle construction appears to be the most popular in Europe (even though it is some times referred to as 'American Projection'). The wide use of the first angle construction (known as 'English Projection') in the United States seems to complete the confusion!

STANDARDS

Drawing standards such as BS308 do more than recommend forms of drawing layout. They suggest the weight and form of lines to be used to depict various features within the construction. Visible outlines and edges should be (relatively) thick and continuous. Inner and leader lines are to be thin and continuous. Hidden features should be shown in short dashed lines whilst centre lines are in thin chains (that is a sequence of long and short dashes). The form of construction to be used for auxilary and sectional views is also laid down.

The codes of practice also go on to cover styles of text and methods of dimensioning features. In order both to save time and to clarify details, symbolic representations can be used for common items. It is, for example, not necessary to draw all the details of the threads on a bolt as these can be represented conventionally by two sets of parallel lines (figure 5.4). Similar symbolic constructions exist for such features as springs, keyways, flats, and knurling. The conventions also lay down good practices which allow effort to be saved; by showing only half of a symmetrical object for instance, or indicating only one of a group of objects (such as one tapped hole on a flange with the remainder being indicated by only their centre lines).

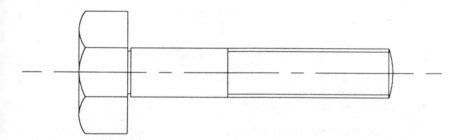

Figure 5.4   Two set of parallel lines used as a symbolic representation of the thread on a bolt.

The engineering drawing must also carry information indicating the materials to be used and the finishes and manufacturing tolerances to be applied to each feature. Various standards have been developed for the description of geometric tolerances. These are mostly based on the work of the International Standards Organisation (ISO). These need to be applied correctly to ensure that the component functions whilst not being unnecessarily difficult and costly to manufacture.

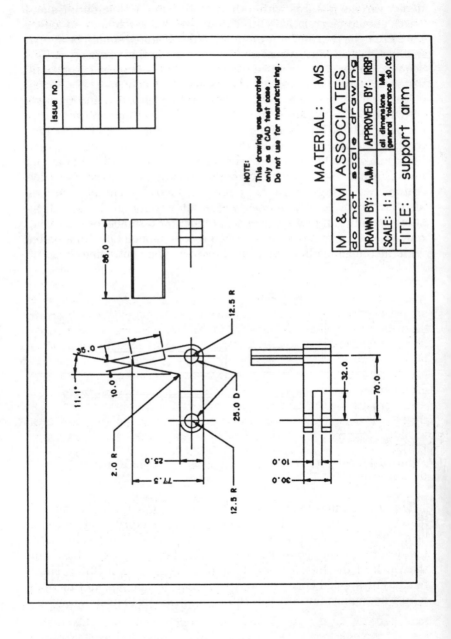

Figure 5.5 A full engineering drawing produced on a CAD system.

There are many two-dimensional CAD systems on the market which have been designed to provide the traditional drafting capabilities and so act as 'electronic drawing boards'. Even more systems have full three-dimensional modelling facilities but are, in practice, used primarily for two-dimensional drawing activities. In this drawing mode most CAD systems can be tailored to operate to one of the major national standards (figure 5.5) and many systems allow drawings to be switched from one standard to another (this includes changing all the dimensions from imperial to metric or vice versa).

USE OF CAD

CAD drafting systems take the physical effort, and some of the skill, out of the job of producing a drawing. Using single commands, features can be copied, mirrored, stretched and scaled, thus saving hours of drafting time. Dimensions (including leader lines and arrows) and text can be automatically generated and positioned. This means that a drawing can be created and dimensioned to a very high standard in a reasonably short time and with very little skill. Paradoxically, these benefits are also the cause of the biggest problem associated with the introduction of CAD. It is now possible to create what appears to be a professional engineering drawing while possessing little or no knowledge of engineering principles. Previous generations of draftsmen spent years developing their drafting skills and, in the process, learned a great deal about the engineering of their companies' products. Today, a novice engineer trained in the use of a CAD system can produce what look to be as good, or even better, drawings than a designer with many years of experience in the company.

Some have claimed that the job of producing engineering drawings is 'on its way out'. This argument is based upon the assumption that the use of full three-dimensional modellers and the transfer of geometric information straight through to numerically-controlled machines eliminates the need for paper copies. This is to ignore the fact that, if the staff involved are to have confidence in the various processes, they need to 'see' what is to be done and that there are always occasions when it is more convenient to carry around a sheet of paper than go to a local graphics terminal.

The drawing will not disappear, but it will be required to address differing communications needs and its form is likely to change. Different kinds of drawing will be automatically produced to fulfil specific requirements, and the engineering drawing will no longer be expected to satisfy all engineering requirements simultaneously. Instead, a whole variety of drawings may be generated, each covering the very different needs of people concerned with purchasing, manufacturing operations, assembly or packaging.

MODULE 5.2 THREE-DIMENSIONAL WIREFRAME MODELS

The use of the node and entity lists as described in Module 3.1 leads naturally to what is called a wireframe model. Because entities such as lines and arcs are used to connect the nodes, when the model is displayed it appears as if wires had been strung along the edges of the object and it is these that are being seen. A wireframe model does not have all the information about a designed component, but it does contain enough for a lot of practical design and analysis work.

PROBLEMS OF WIREFRAME REPRESENTATION

If the design becomes large and complicated, the number of entities increases until there may be so many in a particular view that it is confusing to try to interpret what is being seen. This is particularly true if a team of designers is working on a project and there is a need to look at part designs undertaken by other members of the team.

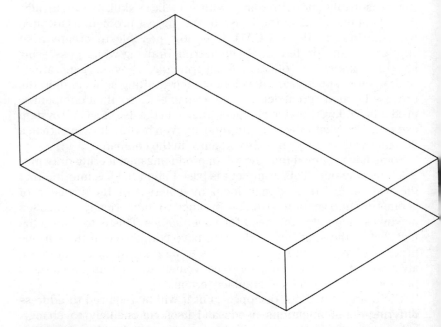

Figure 5.6 The 'Necker cube' illusion – which way round is the matchbox? Is it the complete box, the case or the tray?

Additionally, there is the difficulty presented by the 'Necker cube' illusion. If a cuboid is shown as a wireframe, it is impossible to decide which way it is oriented; indeed the brain may succeed in flipping between two possible situations. The illusion is shown in figure 5.6 which is intended to show a view of a matchbox. One cannot tell, however, if the object is the tray of the box, the cover or the two assembled. In any particular view of a real box, certain edges would be wholly or partly obscured by some of the faces. When the hidden edges are removed from the picture, a better idea of the shape is presented; some examples for the matchbox are shown in figure 5.7.

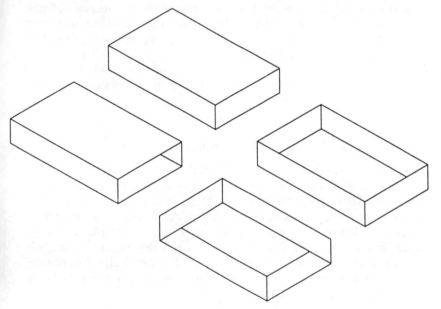

Figure 5.7   Views of the matchbox cuboid with hidden lines removed.

HIDDEN LINE REMOVAL

This process of 'hidden line removal' is a useful one for presenting a better idea of a three-dimensional shape. It must be borne in mind, however, that by removing the relevant edges, information is being removed from the drawing. That information may be vital to a full understanding of the design.

Hidden line removal may be undertaken in two ways. The crude way is for the designer simply to use the 'delete' command to remove edges that should not appear in a particular view. It is, of course, vital to make sure that the edges in question are not removed from the stored computer model itself, thus destroying the design completely. With some CAD systems, it is impossible to delete entities from one view without altering the model; but other systems do allow the user to make alterations to a particular view without changing the model itself. In effect, the view is treated as a two-dimensional drawing and alterations to it are not passed back to the three-dimensional reference data. Even so, care is still needed as removing edges from a view in this way means that the various views of the model are no longer consistent with each other.

The more sophisticated approach to hidden line removal is to have the CAD system perform it automatically on request. In order to do this, the system needs further information about the component. It has to know where 'faces' exist between the edges. In the case of the matchbox shape shown in figures 5.1 and 5.2, for example, more lines are hidden if a full cuboid with all six faces present is represented rather than the cover of the box which has only four faces. Faces can be added to a wireframe model by the use of surface patches of some form. This brings us to the concept of the surface modeller and to the next module.

MODULE 5.3 SURFACE MODELLING

The addition of surface patches to form the 'faces' between the edges of a wireframe model is an attempt to put more information about the design into the computer model. CAD systems which can include surfaces within the description of an object in this way are called surface modellers. The surfaces used need not be complex. A lot of design work is done using nothing but straight line segments. In such cases planar facets are all that is required to make the description more complete. Some CAD systems for architectural applications work on this principle (figure 5.8).

USE OF SURFACE PATCHES

A more sophisticated CAD system may allow the inclusion of surface patches from standard shapes such as spheres and circular cylinders. The latter can, for example, be used to describe areas of blending between two planar regions. Precisely how these are held within the system itself is not of immediate concern to the user. They may be treated as special entity types. Alternatively, such surface shapes may be held in some generic form such as Bézier or B-spline surface patches (as introduced in Module 3.5). This is often the case with CAD systems which allow the user direct access to these types of patch.

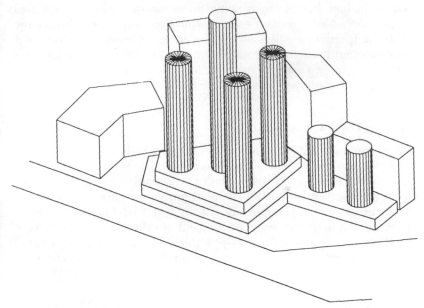

Figure 5.8   Surfaces made up of planar facets used to model a 'building-like' structure.

The user is not normally expected to have to deal directly with the various control vectors that define these latter types of patch. (The option does exist in some systems and is used mainly for the fine-tuning of a surface shape.) In most cases, as mentioned in Module 3.5, the system provides the means whereby surfaces can be constructed by linking or manipulating simpler entities such as curves. The techniques used include the insertion of ruled surfaces between pairs of curves and the extrusion and sweeping of curves around contours. Another approach is to create the boundary of a patch in terms of four curves whose end-points are coincident in pairs and then to allow the system to generate a patch between them. This is not always successful as the result depends heavily on the algorithm implanted in the system to perform the construction. There may be a tendency to produce a flatter surface shape than the user had intended.

FACE LISTS

In order to store information about the inserted surfaces some CAD systems make use of a face list in addition to the node and entity lists. This is not strictly necessary as the surfaces can be included within the entity list provided sufficient space is made available to hold the large number of nodes involved in their definition. Each entry in a face list might consist of the sequence of the numbers of the entities that bound the face. In this way we have an hierarchy: entities are defined as sets of nodes, and faces are defined as sets of entities. Some systems (particularly the solid modellers discussed in the next module) find it useful to reproduce some of this information in a different form and may, for example, also hold each entity as the pair of faces which intersect on it, and each node as the set of entities that meet at it.

SUFACES ONLY WHERE NEEDED

When using a surface modeller there is generally no need to insert surfaces in every possible place. This is a time consuming process and if it is necessary then a solid modeller is probably a better design tool for the application in question. Instead surfaces are put into the model at the areas of critical importance. They help to define the geometry more exactly in these areas. They also allow other constructions to be carried out. For example, it is possible to take sections through a surface in different planes and obtain the appropriate curve of intersection.

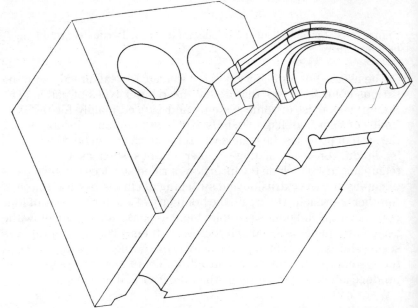

Figure 5.9 Surfaces inserted on critical areas of a wireframe model.

Figure 5.9 shows an example of surfaces being used in this way. It represents part of a punch and die arrangement within a press tool. Surfaces have been inserted around the area of the die cavity and the corresponding forming area on the punch. The rest of the computer model is defined in wireframe. In the course of verifying the design it is necessary to check that the two parts come together in the correct way, that no areas clash and that sufficient space has been left to accommodate the material being formed. If the punch and die are designed on the CAD system as separate models then they can be moved independently on the screen and can be brought together. In the closed position, sections can be taken through both. The presence of surfaces means that there is something for the CAD software to operate on. Various sections can be displayed and the user can check that the punch and die do indeed come together in the correct way.

ADVANTAGES OF SURFACE MODELLERS

As we see in the next module, solid modellers attempt to hold the complete geometric information about an object. Why, then, are surface modellers, which hold only partial information, still important? One reason is economic. Since surface modelling systems do not try to accomplish as much as solid modellers, they tend to be less expensive to purchase. This is often an important consideration for smaller companies wishing to venture into CAD.

Moreover, in order to be able to deal efficiently with the information they hold, solid modellers are often restrictive when it comes to the use of advanced entity types. This may make them unsuitable for applications areas such as the aircraft industry and the design of car body shells which demand the ability to handle complex surface geometries for functional and aesthetic reasons. Surface modellers, on the other hand, have sufficient sophistication in their surface types to allow these requirements to be satisfied.

MODULE 5.4 SOLID MODELLING

As mentioned in the previous module, a solid modelling CAD system is intended to hold a complete geometric description of the object being designed. In other words, as the name suggests, the software should know which parts of the design space contain solid material and which parts are empty. Naturally the amount of data that needs to be held is very large and the calculations required to manipulate it are complex. However the speed of modern computing equipment allows a reasonable response time to user commands to be provided.

VOLUMETRIC AND OTHER PROPERTIES

With more data available to the system, it becomes possible to do more analysis work on the design. At the simplest level, volumetric properties of the object can be obtained. These properties include such things as the volume of the component, its centre of gravity and its moments of inertia. Surface areas of parts of the body can also be determined. At a more advanced level, it is possible to split the design up by means of a finite element mesh and send the resulting data to a special purpose analysis program (Module 8.1). This might find the stress distribution for given loads on the object or, in the case of a design of a die for plastic injection moulding, the rates of flow and of cooling of the moulded material (Module 8.2).

CONSTRUCTIVE SOLID GEOMETRY

Solid modellers are usually divided into two types. In the first of these, called constructive solid geometry (CSG), the model is built up from standard, predefined 'primitive' shapes held by the system. The primitives usually include truncated cones, rectangular blocks, parts of spheres and toruses, and so on. Figure 5.10 shows some examples. The user can select certain of these shapes, specifying the dimensions of each, and can then combine them to form more complex shapes.

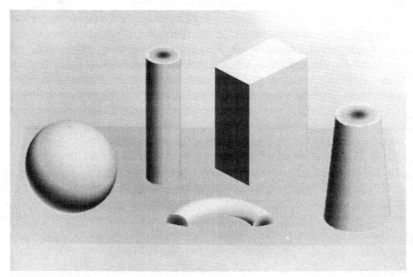

Figure 5.10 A selection of solid primitive shapes.

These combinations are usually formed by means of the Boolean operations of union, intersection and difference. (These are the same operations that are used often in set theory.) Given two shapes, the union is the total volume that lies in both; the intersection is that part (if any) common to both; and a difference is obtained when any part of one shape that lies within the other is removed from the second shape. Figure 5.11 illustrates the operations for two simple shapes.

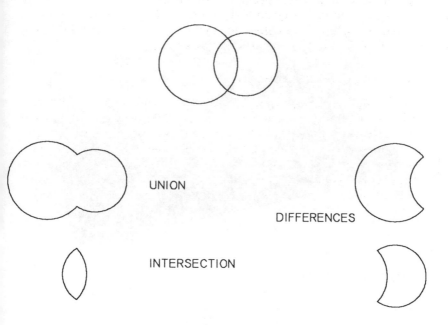

UNION

DIFFERENCES

INTERSECTION

Figure 5.11   The Boolean operations of union, intersection and difference applied to two simple shapes.

The user of a purely CSG system is naturally limited by the repertoire of standard primitives that is available. A large number of engineering components are, however, made up out of simple geometric shapes and so the problems caused by this limitation are not severe. Figure 5.12 provides an example of a component made up from primitives in this way.A CSG system can be said to provide a 'natural' way of describing objects. For the brain tends to break down any complex system into its simpler component parts and a CSG system allows the designer to reverse the process and build the complex out of the simple.

The other advantage of CSG, from the point of view of the CAD system itself, is that it provides a very compact way of holding the data about a design; it consists of the sizes of the primitives involved and the information about how these are combined. Figure 5.13 shows how a fairly complicated component can be built up from relatively simple primitive shapes.

Figure 5.12 Object made up of standard primitives.

Figure 5.13 Complex CSG models can be built up out of relatively simple shapes.

BOUNDARY REPRESENTATION

The other approach to solid modelling that is usually identified is boundary representation (B-rep). The name refers more to how the data is held internally than to how it appears to the user. The component is stored as a collection of entities forming the boundaries of the faces, together with information about the surface patches defining those faces. Care is taken to hold the faces in a systematic way so that the system can tell which side is within the component and which is without.

But the rigid distinction between CSG and B-rep modellers often seems to be somewhat artificial. Most geometric modelling systems today appear to use some version of B-rep data storage for their own internal use. For while the CSG form is very economical as far as internal storage is concerned, the amount of computation needed to make any decisions for subequent operations is almost prohibitively large.

The advantage of the B-rep form is that the information is always readily available. Thus many modellers which have a CSG format as far as the user is concerned, in fact convert the model definition into some type of B-rep form for handling within the system. The system needs to be very careful in its manipulations. The B-rep form can be a simple extension of the node and entity list format, but it makes for greater efficiency if some of the information is held in several forms in different lists. Consequently a good deal of 'housekeeping' needs to be carried out to ensure that the information held is consistent.

COMPARISON OF CSG, B-REP AND SURFACE MODELLING APPROACHES

If the distinction between CSG and B-Rep solid modellers is not clear, it might also be asked what the difference is between a B-rep solid modeller and a surface modeller. For there is a sense in which a B-rep modeller can be regarded as a surface modeller in which all the possible faces have been filled in with a suitable surface. The real differences lie in the greater sophistication of the techniques used to hold the data (in its different forms) in a B-rep modeller and, as a consequence, in the amount of error-checking and other manipulations that the system can carry out for itself.

To take an example, consider the formation of the union of the hemisphere and cuboidal block shown in figure 5.14. If we were using a surface modeller then it is a straightforward task to put surfaces around the curved part of the hemisphere and across the faces of the block. When we put them together, however, a part of a plane surface remains under the hemisphere. Thus the surfaces are defined over more than just the outer shell of the union; there is one part surface

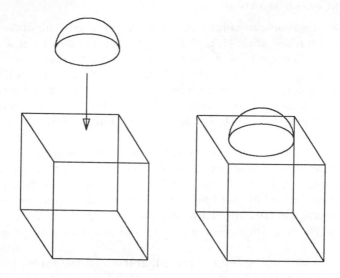

Figure 5.14   The union of a hemisphere and a block.

which is internal to the body where the hemisphere covers the face of the block. It is not easy to use the surface modeller to create a surface with a hole in it for the partly covered plane face. A B-rep solid modeller would give the appearence at least of being able to create the union in the correct way. It may, in fact, still hold the part of the surface under the hemisphere, but it would take care that this was always treated properly.

We can indicate another potential failing of the surface modeller if we now assume that the face of the block to which the hemisphere is to be joined is slightly slanted. As the surfaces are essentially inserted manually by the user, he could well only include surfaces on the block and on the curved part of the hemisphere. Being unaware that the plane face of the hemisphere does not exactly touch the block, the user might leave some parts without a surface so that the insertion of faces for the complete outer shell is left incomplete. With a solid modeller, the outer surfaces of the two shapes would be defined automatically by the system when they were created. When they are brought together by the union operation, this should work correctly to leave a planar surface on the appropriate part of the hemisphere where it misses the block. Thus the more detailed inner working of the solid modelling software produces a more acceptable result.

Surface modellers are, however, more advanced in the forms of surface that they can use. With some solid modelling systems, for example, it is not easy to insert complicated fillet surfaces between primitive shapes. Even when this can be done the new entity cannot always be manipulated with the same ease as the standard shapes. As systems develop, surface and solid modellers will probably move closer and closer together. It will be possible to use surface patches to create new, specialised primitives which can be merged with standard ones in a consistent way.

SPEED OF RESPONSE

As modellers increase in complexity, the speed with which they can operate is reduced. Although improvements in computer hardware help to offset this to some extent, it can still cause problems. In order to overcome these, some solid modelling systems, in addition to holding the precise model of the design, also hold an approximation to it. This is a facetted model in which the various surfaces that comprise the actual form are replaced by a collection of planar facets. Being linear in form, these facets are much easier to handle from a computational point of view and volumes and other volumetric properties can be calculated very easily. Intersections between two models can be evaluated and displayed with speed. Furthermore the approximation is often good enough for most of the design processes. When critical decisions are called for the user can revert to the exact model and be assured of accurate results.

The software associated with solid modellers is very complex. It holds and manipulates a large amount of data, some of which may be duplicated for reasons of computational speed, but all of which must nonetheless be kept totally consistent. The form it displays to the user may not necessarily be the form which is used internally. Indeed there may be more than one internal form — both an exact and an approximate model may be held. Figure 5.15 shows a schematic diagram of a possible solid modelling system.

MODULE 5.5 DISPLAY OF SOLID MODELS

Since solid modellers have a complete knowledge of the component being designed they are capable of producing very sophisticated displays of it. Naturally the greater the sophistication of the display requested, the slower the system is to produce it. The user of a CAD system needs to bear this in mind before making the request. He needs also to think about whether a more elegant type of display will actually assist the design process.

HIDDEN LINE AND SURFACE REMOVAL

All the surface information for the model is in place to define the various faces between the edges. As a consequence the solid modeller can automatically produce a view of the component with its hidden lines removed. There are various strategies that can be adopted by the software to do this. Most are variations on the straightforward idea of comparing every entity against each of the surfaces to decide if it, or any part of it, would be obscured when seen in the current viewing direction. By constructing boxes around the entities and the surfaces (as when intersecting curves in Module 3.5), the rapid elimination of a large number of cases can be achieved.

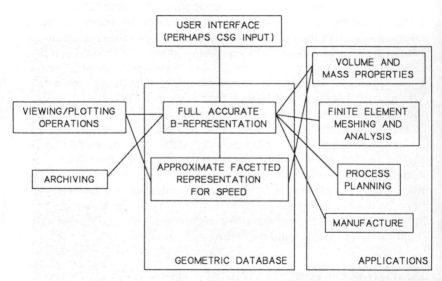

Figure 5.15 The structure of a possible solid modelling system, although it may appear to the user to be a CSG modeller, the internal representations may be held in B-rep form and use may also be made of an approximate faceted representation for some purposes.

If a surface is almost planar, then use can be made of the normal to it pointing away from the main body of the component. If this normal is in a direction which points away from the eye then the surface cannot be seen and so the edges that surround it need not be displayed. As well as eliminating from the display some of the entities that make up the object, it is also necessary to add in some new ones. The most obvious examples are the 'horizon lines' on curved surfaces. If one part of a surface points towards the observer and another part turns away, then the eye should perceive an edge where the surface starts to curve away. This edge is not part of the model and is entirely dependent upon the direction in which the view is taken.

The advantage of performing hidden line removal with a solid modeller is that it is carried out automatically and the edges 'removed' are only eliminated from the display. The stored computer model itself is not affected and remains a true representation of the designed component

SURFACE SHADING

Surface shading is also possible using a surface modeller. This allows very impressive pictures of the component to be produced at the CAD terminal - so impressive, indeed, that they could be mistaken for real objects. An example is shown in figure 5.16.

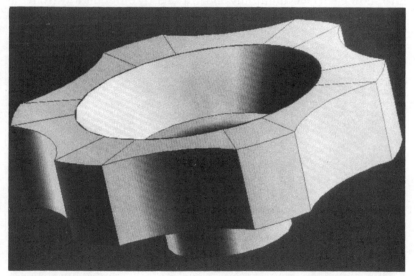

Figure 5.16   A surface shaded image can provide a high degree of realism.

The effect is achieved by using different levels of intensity and shading at the individual points on a surface. To decide on these, the user can place an imaginary light source at any position in the space of the model. The picture is usually built up on a pixel-by-pixel basis. The pixels (picture cells) are the individual points of the graphics screen. The software takes each pixel in turn and first decides if it is part of any surface within the model that is visible along the current viewing direction. If it is, then the normal to the surface at that point is found. The orientations of this to the viewing direction and to the line between the surface and the light source are used to determine what intensity and shade should be given to the pixel. The colour itself is usually selected by the user who attributes a single colour to each surface (or to each primitive shape) within the model. If the pixel is on a surface which faces away from the light source, it is shown more darkly and with less intensity than if the surface faces the light.

There are hundreds of thousand of pixels on a good, high-resolution graphics screen and the amount of computation needed to generate this type of shaded image is very large. Again, the process can be speeded up by using a facetted approximation to the model. In this case, each facet is taken in turn. The software then decides if it is visible to the user along the viewing direction (the idea of the outward pointing normal is useful here). If a part of it is visible then the whole of that part is shown with the same intensity and shade. These are again determined from the orientation of the facet to the light source and to the viewing direction. While this is a quicker process, the effect is to show each facet uniformly and in a slightly different way from the adjacent ones. Thus the user can see the individual facets and this detracts somewhat from the overall effect (figure 5.17).

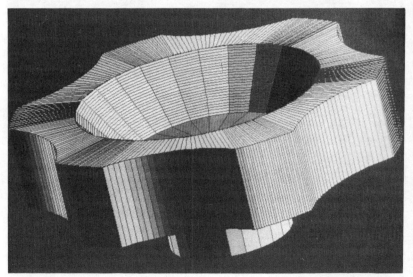

Figure 5.17   A faceted image.

NEED FOR HIGH QUALITY SHADED IMAGES

It is questionable if the ability to produce high quality shaded images is useful as part of the design process. The time taken to generate them is large, and even if the comparatively speedy facetted form is used, the information it conveys is no greater than the original wireframe view (if this has been well laid out). Indeed hidden line and surface details have deliberately been obscured. These may well be of prime interest to the designer in deciding if the component is correct and will function correctly. Good quality images do have a place in the preparation of technical documentation and for publicity work. Some companies, indeed, regard the ability to show glossy, shaded images of their products taken from their CAD systems as an integral part of their sales strategy. Good exploded views of a complicated assembly with hidden details removed are an excellent way of showing how the parts should fit together. These can then be used on the shopfloor when a product is being produced; they can also be incorporated in the requisite maintenance literature for that product.

EXERCISES

1. Select a number of items with differing degrees of symmetry and sketch the minimum number of views necessary to fully describe these objects. Identify those cases where three views are required and those where all information can be carried in a single view.
2. Look up, in BS308 (or in student publication PD7308), the representations used for knurling, flats and thread form. Produce an engineering drawing that includes all these features. Also add major dimensions to your finished drawing.
3. Comment on the changing role of the engineering drawing. Suggest ways in which the various new objectives can be met automatically upon a CAD system.
4. 'But will it produce ordinary engineering drawings for me?' How far do you think this question, which is often posed to CAD vendors, is a valid one?
5. What type of modelling system (if any) would you recommend for companies producing the following items?

   press tooling for car body panels
   office furniture
   designs for office layouts
   moulds for plastic bottles
   hot forging dies for connecting rods
   production line machinery for food processing

6. What are the reasons for holding a facetted approximation to a full solid model of a component? What are the potential disadvantages of such an approach?
7. It has been argued that the ability to produce shaded images is irrelevant to engineering design. The image only serves to hide

internal details which may be of critical importance. This view might also suggest that shading capabilities are only incorporated in advanced CAD systems to impress the managing director of a company which is a prospective purchaser! What advantages do shaded image pictures offer to the whole of the engineering design process?

SECTION 6
THE USER INTERFACE

A CAD system is a tool for the designer. If he is to make full use of the system, the user must be able to interact easily with it. This interaction takes place via the user interface, the facilities such as the keyboard and graphics tablet (see Module 2.1) that allow the user to input information to the system.

Information is usually put into a CAD system by means of a range of predefined commands. These form the 'command language', and the designer will have to master this language before he can use the system. The commands are either entered via the keyboard or are selected from menus on the graphics screen itself or on the tablet. The user selects a command from a menu by pointing to a designated area on the screen or tablet. (Some attempts are being made at direct voice input of commands, but this is still in its early stages as far as commercial CAD systems are concerned and this technique is not dealt with here.)

Each command instructs the system to perform one action and, in some cases, a sequence of commands may be used to define each of the standard operations. In addition, the user may be invited to specify a few parameters which effectively define the action to be taken. This brings us to the area of parametric programming where standard commands, with critical values replaced by parameter names, are held in a disk file and called up when needed by the user who then specifies the actual value to be used for each parameter.

There is, however, a limit to what can be accomplished in this way and many CAD systems are equipped with a 'graphics interface language'. This is a programming language in its own right which can interact with the geometric and other databases of the CAD system so that information can be put in and extracted.

A CAD system is sold as a general-purpose package capable of handling a wide range of engineering problems. Many of its features may well not be used by all the companies which buy it. But to realize the full benefits of the system, it should be tailored to the needs of those using it so that the types of operation required most frequently are easily accessible. This tailoring may be simply a question of creating a special-purpose menu incorporating those commands that are found to be required frequently. It may involve parametric programming, perhaps using the interface language, so that complex but widely used procedures can be handled easily. The tailoring process may by itself be time-consuming, but unless it is thought about carefully and undertaken sensibly, the full benefits of CAD techniques will not be obtained.

MODULE 6.1 USER COMMAND LANGUAGE

For a computer program to be able to perform more than one specific task, it needs input from the user. This input determines what output function is to be generated.

Numeric Input

Very simple computer programs take numeric data from the user and employ it to produce numeric output. Such programs may provide a useful design aid. To give a very simple example, a program could be written to investigate the layout of a number of holes equally spaced around a given pitch radius in a circular plate. When the program is run, the user would need to provide information such as the diameters of the plate and the holes, the number of holes, and the required pitch radius.

If the program were 'user-friendly', it would perhaps display lines of text on the screen to tell the user what information should be entered next. Once it had the data, the program could perform the various trigonometric calculations and produce an output, which might be a hardcopy plot of the layout of the holes. (A more sophisticated example of a design program relying purely on numeric input is a finite element analysis package whose basic input is the geometry and topology of the nodes and elements of the structure and various properties of materials.)

Command-driven Programs

If a design system is to be flexible, it needs to be much more interactive. This is particularly so in the case of the CAD system. The nature of such a system is that it is open-ended. A number of capabilities are provided and it is up to the user to tell the system what use he wishes to make of them. The architecture of the CAD system itself comprises a number of procedures, or subroutines, each of which provides one (or sometimes more) of the individual capabilities or functions of the system. For example, one such procedure may be used to insert a line into the geometric database; another procedure might then be used to display it.

In order to tell the system which function is required, the user needs to input a command. On a very simple CAD system this might be a single letter, perhaps 'l' to insert a line. The system then enters the appropriate procedure to implement the command. Within the system program itself, this is most easily achieved by use of a programming construction like the Pascal case statement or the FORTRAN computed goto statement. The procedure chosen may itself require extra information. In the case of inserting a line, for example, the system must be told the positions of the end points, and the procedure will therefore incorporate prompts which invite the user to supply this information.

Verb-noun Commands

This type of approach has led, over the course of time, to the development of user interface languages based upon 'verb-noun' commands. The typical form of such a command is:

<verb> <noun> <qualifiers>

The verb is simply the basic command telling the system what function is to be performed. It may exist on its own. For example, a common requirement for a CAD system is to clear the screen and redraw the display. This might be invoked with the command:

REPAINT

or an abbreviation, say REPA. (This and the other examples used in this section are actual commands available at the time of writing on the Computervision Personal Designer system, but very similar commands are available on other CAD systems.)

The noun part is required when the command can act in several different ways. It helps to specify the user's requirements more exactly. An example is the command:

ZOOM ALL

which would be used to instruct the system to clear the screen and redraw all parts of the current design at a scale to fit onto the graphics screen.

Additional Qualifiers

The qualifiers provide additional information, usually in the form of numerical values. For example the command to insert a line between the points with two-dimensional coordinates (1, 2) and (3.5, 4) might be:

INS LIN: X1 Y2, X3.5Y 4

(In this form of the command, the colon serves to separate the main body from the positional qualifiers that apply to it, and the comma to separate the coordinate values.)

At first sight this form of command may seem rather cumbersome to use, and while the form exists on most CAD systems, attempts are usually made to ease the burden on the user. One approach is to use the graphics tablet as an input device to help in the issuing of a command (see Module 2.1) and in the selection of geometric information; this is discussed in the next module.

COMMAND INTERPRETATION

We conclude this module by noting that when designing a CAD system a major effort is required to find ways in which even the very simple form of command line given above can be handled. The command comprises both text and numbers. As a consequence the whole line entered by the user has to be handled by the system as a character string. In order to identify the command itself, the first part of the string has to be extracted. Thus the system has to search the string for the first space, ignoring any leading spaces before the command verb itself. Once the command has been identified the first level of branching to the appropriate procedure can be done. Within that procedure, the next command word (the noun) has to be extracted and branching is again required. The qualifiers may be a mixture of letters and numbers. If numbers are present, each has to be reformed as a real or integer number from the characters that the user has typed in.

There is, therefore, a large computational overhead involved in making a CAD system easy to use. This overhead is demanding both in terms of the use of the computer and its memory space, and in terms of the suppliers' development and maintenance costs. A good user interface is every bit as important as sophistication in the handling of graphics and geometry.

MODULE 6.2 USE OF MENUS

The command language described in the previous module is a very powerful way of interacting with a CAD system. It does, however, have the disadvantage that typing a sequence of complex commands on a keyboard quickly becomes wearisome and some would also argue that the keyboard is not in any case a very efficient means of entering data to a computer system. To overcome these objections a number of different ways of issuing CAD commands have been devised.

Most of these depend on the use of menus of one form or another. These comprise commonly used versions of the commands each of which can be invoked by a single action on the part of the user.

MENUS ON THE TABLET

A popular way of allowing the user access to a menu of commands is to make use of part of the graphics tablet (see Module 2.1) which is also used to input geometric positional data. The part of the tablet area reserved for the menu is divided into a number of rectangular areas. These areas are actually defined to the computer in terms of the coordinates of their corners. But in order that the user can identify and locate them a paper or plastic overlay divided into rectangles is put on top of the tablet. (Naturally it is important that the overlay is positioned correctly on the tablet so that the areas on the overlay align with those it is identifying on the tablet.)

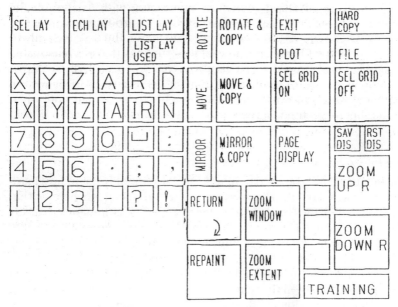

Figure 6.1   Part of the menu area from a graphics tablet.

Associated with each area on the tablet there is a string of characters. This may be either a complete command or it may represent a part of a command. When the user selects an area by pointing to it with the pen, the associated piece of text is retrieved and displayed on the screen. If it is a full command, it is obeyed immediately. If it is only part of a command, the system waits for the user to complete it by making an entry via the keyboard, by selecting geometric information (using the tablet) or by pointing to other menu areas. The use of a tablet menu thus enables the user to issue a command more quickly. So that it is clear which area represents which function, text can be written into the appropriate region on the overlay or, more compactly, a small pictorial representation can be drawn. Figure 6.1 shows an example of part of the menu area from a tablet overlay.

USER CONFIGURATION OF THE MENU

An advantage of this form of menu is that the user can easily configure it to meet his own needs; he can rearrange the menu so that the commands which are commonly used for his particular application are present and others are removed. The latter however can still be invoked by entering them directly via the keyboard.

A menu of this type is held within the system as a file that associates the appropriate character string with each menu area (defined by its corners). By associating each character string with a label on the menu it is possible to get the CAD system to produce a new menu overlay when required. To reconfigure the menu, the user has only to edit the defining file and, because of its structure, this is a reasonable straightforward procedure. It is also possible to change from one predefined menu to another. All that is necessary is to provide a command which allows the second menu file to be read in, and, when this has been executed, the user substitutes the new overlay for the old one.

SPECIAL SYMBOLS

The menu can also be used to insert special symbols or standard parts into a design layout. These symbols or parts can be held within libraries in disk files on the computer system. The commands to insert these into a design can then be incorporated in the menu. This facility is particularly useful in the laying out of electrical circuits. The library contains simple drawings of the individual electrical component symbols and areas of the menu are marked with these. The user can call these up, position them and then complete the circuit by drawing in the interconnections with simple geometry such as lines and arcs.

FUNCTION BOXES

An alternative to the tablet-based menu is the 'function box'. This is effectively a special keyboard. Each key is associated with a particular command or part command in precisely the same way as the areas of the tablet. As the number of keys is limited (often to nine) the association between the keys and the commands has to be changed more frequently. Partly for this reason, and partly because the tablet is a more general purpose piece of equipment, the function box does not seem to have gained great popularity.

ON-SCREEN MENUS

A disadvantage of the tablet-based menu and of the function box is that the user must continually transfer his attention from the menu to the graphics screen and back again. To overcome this problem an alternative method of presenting menus has been tried. This involves using areas of the screen to display the commands. Usually the areas designated for the menus are around the edges of the screen. The user selects one option by moving the screen cursor to the relevant area (and then pressing an appropriate key on the keyboard or mouse being used). Swapping from one menu to another is easier than with the tablet as the system itself redraws the areas with the new symbols to form the relevant 'overlay'. Modern workstations are using this approach to facilitate access to common, general-purpose commands. The areas themselves may not be specifically delineated but are identified by the symbol itself and it is this that needs to be 'hit' by the user. These symbols can be relatively complex (given the sophistication of the graphics on these workstations) and the current terminology refers to them as 'icons'.

Because screen space in any CAD system will always be at a premium, the use of the kind of on-screen menus described above has obvious disadvantages. (This is perhaps less so now that screen sizes are becoming greater.) To avoid this difficulty, a different kind of screen menu has been developed. The user selects the desired function by running through a sequence of menus; at each stage, a small list of options - usually ten or less - is presented to him. Each option represents either a single function to be performed or else an indication of the type of command sought. The user selects an option by entering its number or by pointing at it using a light pen or mouse. If this specifies only an indication of the type of commands, the system then presents a more refined list of options that relate to the area previously specified by the user. In order to invoke a command the user thus works his way through a hierarchy of options until the precise action required has been defined. Figure 6.2 shows a part of this type of hierarchy schematically.

This approach may seem somewhat cumbersome to use although it has been found that, with practice, the sequence of options needed to reach commonly used commands can be memorized. But this does take some time and the casual user of the system would not gain this type of familiarity. Another disadvantage is that is is not easy to allow the user access to reconfigure the order and composition of the option lists to his own particular needs. Purchasers of such systems are therefore constrained, in that they always have to work in the way in which the system designers thought was most appropriate.

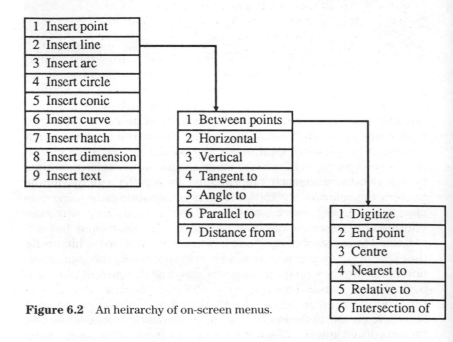

Figure 6.2 An heirarchy of on-screen menus.

PULL-DOWN MENUS

Nonetheless on-screen menus of this latter type are being used on many modern workstations for common general-purpose functions. By indicating a particular point on the screen a predefined 'pull-down' menu can be made to appear as though a roller blind had been pulled down to cover a small part of the screen. This menu lists functions that can be performed and the user moves to the appropriate one with the cursor. (Nesting of lists is not common in this usage.) Once the selection is made, the menu is 'rolled up' again, that is it disappears from the screen and the part of the image it replaced is reinstated. There is, naturally, a computing overhead associated with this type of facility and it is only really practical using workstations which have their own processing capability.

As has been seen, menus have developed as a means of easing the task of entering commands into CAD systems. Various different approaches have been used. Their success depends to some extent upon who is using them. It is very difficult to design an entry procedure which is easy for the casual user and yet does not limit the speed of interaction for the person who is familiar with the system.

MODULE 6.3 GRAPHICS INTERFACE LANGUAGES

Graphics programming started to come into its own in the mid 1960s. It was then that programs began to be developed which, as well as acting as pure 'number crunchers', also provided their output in graphical form so that it could be easily comprehended by the user. Such programs could indeed be used as aids to design, but they did not constitute CAD as we now know it.

GRAPHICS LIBRARIES

In order to output graphical results, a program needs to give instructions to the graphics output device. This is done by transmitting sequences of characters which are encoded commands. These are interpreted by the output device which then takes the appropriate action. The instructions are often to move the pen (for a plotter) or the cursor (for a graphics screen) to a new point, perhaps at the same time drawing a line. Other commands cause the device to be put into the correct mode to start a plot or terminate plotting altogether.

It would obviously be tedious for a program developer to generate the appropriate coded instructions every time they were required, and libraries of routines to perform standard graphical operations were therefore developed. By a single call to the appropriate routine, the program writer could initiate the graphics mode. Another call would allow a line to be plotted to a new point. Coordinates could be given in the coordinate system most natural to the application and the routine would itself transform them into coordinates suitable for the output device. This approach eased the burden of program development. It also made any particular piece of software less dependent upon a particular hardware configuration, since it could be transferred to any other machine upon which the graphics library was already installed.

These types of graphics libraries are still in use today for this type of application. They are called from a user program written in a high-level language such as FORTRAN or C. Examples of such libraries are GINO, GHOST and SIMPLEPLOT. Attempts have been made to standardize in this area and one result has been the emergence of GKS, the Graphics Kernel System, which comprises a list of prescribed, named subroutines that perform specific graphical functions. The intention is to facilitate the portability of software.

All modern micro-computers have graphics capabilities. Often they are supplied with the programming language BASIC. The user can access the graphics by use of special purpose programming statements that have been added to the BASIC language itself. For example, many implementations have 'move' and 'plot' statements. These provide the same facilities as a library of routines but do so differently, in that they are built into the programming language itself.

Using these types of software tools, a program writer can create his own programs which hold and manipulate data and which are then capable of producing graphical results. A CAD system is an example of such a program. It is written in a high-level language (often FORTRAN or C); it holds geometric information (in node and edge lists); it can process that information and output it to a screen or plotter.

Need for Tailoring a CAD System

A CAD system provides a wide range of commands and processing capabilities. It is a general-purpose piece of software which the vendors have designed to meet the requirements of a large number of differing applications. This means that it may not handle some intricate task peculiar to one particular application very efficiently. There is, therefore, a need for the user to be able to tailor the software to meet his own needs.

Use of Macros of Commands

To start at the simplest level, suppose that an application requires a user to go through the same sequence of commands many times during the design process. His job can be greatly simplified by placing that sequence of commands into a disk file on the computer system, so furnishing a 'macro' which can be called up every time the sequence is required. The macro could, for example, be invoked by the selection of a region on the tablet menu.

Graphics Interface Languages

However, the application may require different actions to be taken depending upon different conditions. It may, for example, demand that teeth be laid out around a gear wheel. The output depends on the number and size of the teeth and also on the radius of the wheel. Here a simple sequence of pre-defined commands does not suffice. Some decision-making and calculation is also required. To handle this situation, CAD systems often provide a graphics interface language. This is a type of high level language that allows the user to gain access to the geometric data held by the system and to some of the geometric manipulation procedures built into it. This is an extension of the idea of using a standard programming language and one of the libraries of graphics routines. The CAD system itself is now acting as a source of geometric procedures.

The form of the interface language varies from one CAD system to another. Most try to look like one of BASIC, FORTRAN or Pascal. Indeed, in some instances, the language is precisely the standard one and the calls it makes to the CAD system are ordinary library calls to

```
-------- Demonstration Program ------- BOOK.UPL ----
----------------------------------------------------
--   sets a number of holes around a central one.
----------------------------------------------------
---------------------------------- MAIN PROGRAM ----
PROC MAIN
INTEGER N,I
REAL SRAD,PRAD,PCD,NANG,PANG
COORD SC,PC
WINDOW 9,28,32,1,40
---------------------------------- clear system ----
PRINT_WIN=9
CLEAR 9
ACCPT_WIN=9
------------------------------------ entry data ----
PRINT"...CENTRE HOLE..."
ACCEPT SC.X PROMPT('... x-position..')NEWLINE
ACCEPT SC.Y PROMPT('... y-position..')NEWLINE
ACCEPT SRAD PROMPT('... radius..')NEWLINE
ACCEPT N PROMPT('...PLANET HOLES..number...')NEWLINE
ACCEPT PRAD PROMPT('    ... radius..')NEWLINE
--------------- calculate hole angle and radius ----
NANG=360.0/REAL(N)
PC.X=SC.X
PC.Y=SC.Y+SRAD+2.0*PRAD
-------------- draw main hole and top aux. hole ----
SEND
SEND 'INS CIR R ',SRAD,':X',SC.X,'Y',SC.Y
SEND 'INS CIR R ',PRAD,':X',PC.X,'Y',PC.Y
PC.Y=PC.Y+PRAD
N=N-1
--------- copy top aux. hole to other positions ----
LOOP I = 1 TO N
PANG=NANG*REAL(I)
SEND 'ROT COPY ANG ',PANG,':X',PC.X,'Y',PC.Y,',,X',SC.X,'Y',SC.Y
END_LOOP
PRINT"....construction completed....."
------------------ reset CAD to normal graphics ----
CLEAR
WINDOW 10,1,4,1,80
WINDOW 10
PRINT_WIN=10
ACCPT_WIN=10
END PROC
----------------------------------------- END ----
```

Figure 6.3 A program written in UPL that positions a number of holes around a central hole.

the procedures from which the CAD system itself is built; this is a very flexible approach. The amount of interaction with the CAD database and graphics also varies. In the early days this access was deliberately limited, resulting in closed systems which could not easily be tailored to specific applications. More recently, systems have become more open and attempts are now being made to produce standard means of access for all systems. Quite how much success these will enjoy remains to be seen — there are good commercial reasons why the CAD vendors are not keen to have user software and data too independent of the system upon which it was generated.

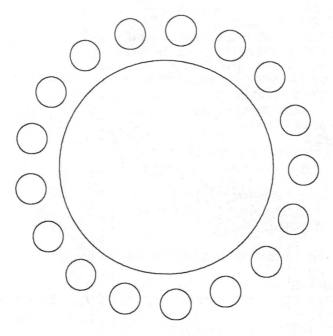

Figure 6.4 The outcome of a trial run of the UPL program shown in figure 6.3.

Graphics interface languages are provided so that CAD systems can be tailored to perform standard operations associated with specific applications more easily. As they resemble high level languages, some programming experience is required before they can be used successfully. Consequently many general users do not make use of them. In such cases programs have to be developed in the user company by computer system personnel, who must, of course, liaise closely with the design staff to produce a usable piece of software.

Two Examples

We end this module with two examples. The first uses the User Programming Language (UPL) of Computervision's Personal Designer system. Figure 6.3 shows a UPL program for positioning a number of holes equally spaced around a central one. Here the central hole is first specified and located. The number and size of the 'planetary' holes are then requested. The program can then calculate their angular positions and pitch radius (this being chosen to make the gap between the planetary and central holes equal to the radius of the planetary holes). The exact position of each hole is thus determined and they are drawn starting with the first directly above the central hole. Figure 6.4 shows the results of a trial run.

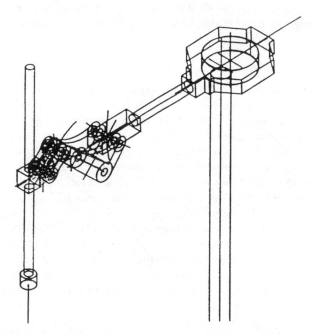

Figure 6.5 Simulating the operation of a mechanism using a program developed with the CADDS3 software.

The second example is more complex. It was developed using Computervision's CADDS3 software which is based on a mini-computer. The results are shown in figure 6.5. This shows a simulation of the operation of an industrial sewing machine. Each part was created on the

CAD system as an individual model and each is stored in a separate file on the system.

The program written in the graphics interface language controls how these parts are put together on the graphics screen. It undertakes the calculations necessary to ensure that the parts stay together in the correct configuration as the input link is rotated. Here the interface program performs geometric calculations based on the required transformations. It also holds the various positions and, from these, performs numerical differentiation to determine the velocities and accelerations of the individual links. From this information it can proceed to plot velocity and acceleration diagrams which help the user to identify possible problem areas; for example, positions where interaction forces are large.

MODULE 6.4 USE OF PARAMETRICS

This module is a continuation of the previous one, in that a graphics interface language is the ideal way to carry out parametric programming. Let us begin by looking at the sort of ways in which parametric programs are used.

PARAMETERS IN DESIGN RULES

Very often the problems in particular design areas are well understood and a new design can be created easily by applying certain standard design rules. This may sometimes involve a certain amount of calculation in order to ensure that critical factors are taken into account, or it may simply be a question of reading standard dimensions from a data sheet. A very simple example might be the problem of determining the dimensions for a bracket to support a load. The thickness, length and width are critical to ensure that the bracket is strong enough.

A table of the standard sections to be used for every range of load cases likely to be encountered may be available to the designer, in which case the appropriate values can simply read off and the bracket drawn accordingly. More complicated examples are the laying out of gear teeth around a given circumference, where the important values are the number of teeth and dimensions giving the size of each, or the arrangement of standard support blocks within a press to hold the part within the actual die cavity itself, the blocks here are available in standard sizes.

USE OF THE GRAPHICS INTERFACE LANGUAGE

This type of problem can be handled most conveniently by programs written in the graphics language of the CAD system. Such a program prompts the user to input the critical values it requires. Once these have been provided, it can proceed to perform any necessary calculations, and can then go on to construct the necessary geometry and display it on the screen. However, someone needs to write and test the graphics program itself. This requires a certain level of expertise in computing as well as a thorough understanding of the engineering problem being tackled. The ordinary user of the CAD system may not possess the required level of computing expertise or, indeed, have the time available to develop the program properly.

USE OF THE COMMAND LANGUAGE

Because of the degree of skill demanded by the interface language, a number of systems allow easier access to parametric programming for problems of a simple nature. Often such access is gained via the user commands themselves, with which the ordinary user can be expected to be familiar. The commands are normally given via the keyboard or one of the menu configurations. However, since they are simple text strings, there is no reason why they could not be read off a text file held in the disk filing system of the computer. If such a file is created containing ordinary CAD commands then, when it is invoked, these commands are obeyed by the system. One could, for example, use this technique to create a drawing border around a design layout (although it would be quicker merely to store the border as a standard part and insert it wholesale when required).

Such a file performs the same action every time it is used because the same commands with the same numeric values are always present. However, if these values were replaced by symbolic variables (as in a programming language), the result can be made to vary. When the file of commands is now run, the system encounters names which ought to be numbers. It can recognize them as such and can prompt the user to supply the necessary values if these have not previously been given. This approach allows the user to create a parametric program very easily. The usual CAD commands are used with symbols replacing critical variable dimensions and parameters. If algebraic expressions can be substituted for the symbolic names, then even greater scope is available. There is, however, a limit to what can be achieved, for it is difficult to incorporate loops and conditional statements into this type of structure. Thus little real decision-making can be undertaken.

USE OF A JOURNAL FILE

As an alternative to creating the file of commands directly, using an editor, some systems allow it to be generated while the CAD system is being used. In order to do this a 'journal file' is kept. This contains a copy of some or all of the commands issued by the user during a session at the CAD workstation. This naturally contains the values given to each command as well. Thus the user can create an example of the parametric part and have the procedure recorded at the same time. As an aside, we should note that the journal file is also used by some systems as a protection against the effects of the system failing while in use. If the commands issued have been held in a file then they can by 'replayed' once the system is up and running again. Only those commands issued at the point when the failure occurred are irretrievably lost.

Once the example of the parametric part has been designed the journal file contains the commands to generate that particular component. Before other examples, with changed parameters, can be created it is necessary to edit the file, removing extraneous commands and changing specific values into symbolic names. However as the commands have already been used successfully, part of the testing procedure is already complete.

SUMMARY

Parametrics are used for specific design problems which are well understood and which rely on relatively few critical values. A parametric program can be generated as a sequence of user commands held in a file containing symbolic names rather than actual numbers. The use of parametrics can speed up the preparation of a drawing, particularly if a lot of repetition is involved. They can also be used to encapsulate the expertise of an experienced person in a particular area so that it becomes more generally available to other designers. More elaborate programs require the use of a full graphics interface language.

EXERCISES

1. If you have access to a CAD system, investigate the form of its command language. How does it compare with the 'verb-noun-qualifier' structure described in this section? How easy is it to learn this structure, and to use it, and to what extent is it the best form available?

2. What are the relative advantages and disadvantages of using a menu on a tablet, on the screen, and on a special 'function box'?

3. What are the drawbacks of using a full graphics interface language? Discuss who within a company should be trained to use such a language and who should be responsible for the maintenance of the programs generated?

4. If you have access to a CAD system, investigate the nature of its interface language or other parametric programming facilities (if any). What form do these take and how easily can they be used by someone without computing expertise?

5. Tailoring a CAD system is important as a way of creating greater benefits. How would you tailor a general-purpose, three-dimensional wireframe modelling system for use in the design of office furniture and/or printed circuit boards?

SECTION 7
SYSTEM EFFECTIVENESS AND ORGANIZATION

The effectiveness of a CAD/CAM system within a given industrial context depends upon a large number of interrelated factors. These can be classified under two main headings: those that bear upon the technical operation of the system and those that relate to its social acceptability. If the facility is to be of benefit then it must perform the necessary design activities and the staff must be committed to using it. Without both the capability and the commitment, success cannot be achieved.

Although any CAD/CAM system has a wide range of technical capabilities, only a limited number of these may be required in order to successfully design and manufacture a given product. At the same time, however, it is vitally important that the system chosen, or currently in use, should be matched as closely as possible to the needs of those who use it. Mistakes made during the selection process can result in a less than ideal system being installed. Similarly, the changing design requirements of the company need to be regularly monitored, lest the company's procedures become less effective because the system is unable to meet new requirements as they are introduced. The capabilities of the system must always remain in step with the current design needs of the business.

It is often suggested that the individual design activities of a company should be determined and compared directly with the capabilities of a CAD system. This would suggest that, having established that a certain number of man-hours are required for manual drafting activities, the effectiveness of any CAD system being considered can be measured simply by determining how many of those man-hours would be saved if the system was introduced. Unfortunately, this kind of approach ignores the far greater gains to be made by producing the information in a different manner so that it can be automatically used in other downstream operations. Far greater cost savings may be achieved by *increasing* drawing office time, and looking at alternatives and considering likely problem areas, than can ever be made by simply producing drawings faster.

It is thus necessary for a company (or the individuals responsible) periodically to review the technical needs of their design process. To do this effectively the flow of technical and commercial information throughout the organization must first be determined (Module 7.1). The resulting information network then indicates those critical points at which information converges and activities are performed. All such activities generate various types of outputs, which in turn lead to other activities that directly or indirectly create the final product. Such outputs may be anything from the purchase of a subcomponent to the definition of a geometric value, or the completion of a manufacturing operation, or the assembly of some components or even a written message indicating that an activity has either been successful or has failed.

The individual design activities can thus be considered as providing some modification, assembly or expansion of the inputs (be they knowledge, the developing scheme or the product itself) that produce a desired output. Once the network has been refined to make good any obvious deficencies, the individual activities can be evaluated to establish the design needs (Module 7.2). By considering the operation to be performed, its effectiveness can be gauged and compared against manual practices. In this manner a true comparison can be made and the areas of greatest benefit established (Module 7.3).

It is a common assumption that it is only necessary to install a CAD system in order to obtain the desired productivity improvements. But many cases can be quoted where this has not happened simply because there was a failure to involve the staff in the operation. The problem may have taken various forms. Perhaps there were insufficient training places; or maybe staff were expected to maintain high productivity while the system was still at the introductory stage (which effectively forced them to continue using manual practices as they did not have time to learn how to handle the new technology). Whatever the particular cause, the lesson should be clear: the involvement of the staff must begin at the earliest possible time. For a new installation this should be as soon as the possibility becomes a matter for investigation. Once a system is installed, the involvement should be maintained by a continuous programme of training and by ensuring the awareness of its use throughout the organization (Module 7.4).

The working environment must be considered (Module 7.5) on a day-to-day basis in order that the trained staff are best able to operate effectively. Again, it has been reported that the staff are upset that while great care is often devoted to providing the computers with the right environment, no one gives a thought to the working conditions of the operators. The small amount of additional effort required to make sure that the right conditions are provided for those who use the system can greatly improve its effectiveness.

Finally, the two themes of technical and social concern must be drawn together so that an appropriate approach to managing the system and controlling the design activities can be introduced (Module 7.6). All too often these processes are reduced simply to ensuring that the system is available and making sure that the fullest possible use is made of all the workstations. The design process, however, needs to be carefully planned to ensure that the necessary (and only the necessary) activities are performed at each stage. Unnecessary work, whilst increasing the system utilization figures, does not help reduce the product development cycle time. Information once entered and checked, must be made available in all other operations, thus reducing not only execution time but also the chances of errors.

The information contained within the system is vital to the successful development of the product design. It must therefore be handled and cared for as though it were 'priceless'. It must be monitored and controlled at all times so that it is available (and uncorrupted) when called for by the designer (Module 7.7). Only when this controlled data is brought to bear upon the appropriate technology, and when this is manipulated creatively by the designer, can a CAD/CAM system be said to be truly effective.

MODULE 7.1 INFORMATION FLOW

In order to understand the design needs of a company it is often easiest to consider the complete organization as though it were a 'black box'. Such an approach allows the important activities to be quickly identified and their influence, one upon another, to be assessed. By using this technique it is possible to establish what kind of CAD/CAM installation will meet the company's needs or, if a system is already in place, to measure its effectiveness.

STIMULATION OF DESIGN ACTIVITIES

Information and resources can be seen to flow into the black box, where they are used to stimulate the 'function modifying processes' (that is the activities being performed within the plant) in order to produce the desired output - the product. Without even knowing what is happening behind the closed doors of the factory it is possible to observe and record everything that enters and leaves. It must be these resources and information that result in the product and the other observed outputs (such as wastage), if the factory is to operate as a stable unit.

RESOURCE BALANCE

If the product outflow contains more resources than are received then, over a period, the factory will be gradually depleted. Someone inside is either stripping out the fittings or they are mining for coal! Conversely, if more resources enter than are detected as leaving then the surplus is either seeping out unnoticed, or the waste is building up within the plant. For the factory to remain productive over any reasonable period a balance must be maintained between the inputs and the outputs of the black box.

MAJOR RESOURCES

The resources entering the premises are at first sight easy to detect. They comprise power, materials and people. It is the power input, together with the skills resident in the people and factory (such as machine tools and work practices) that transform the material into saleable products. The secondary effects, such as the input provided by the workforce, could also be included in this study but is unnecessary as the main aim is to establish the information flowing through and around the factory. It is this information that dictates how and when an activity is performed upon a particular set of resources to ensure that the product goals are achieved. Thus by tracking the major resources it is possible to infer (and then to check) what information is necessary to create the conditions which modify or merge them to produce the end product.

CUSTOMER RELATED INFORMATION

The first and simplest impression is that any factory operates on the basis of a single input instruction, namely an order to supply. This results in the resources of men, machines and materials being used to 'forge' a single output product. But such a model, whilst possibly adequate for describing a store or shop, is too simple for a manufacturing environment.

Information may initially flow between customer and factory, perhaps starting at the tendering stage and continuing, with queries and proposals being exchanged, until a full product specification is agreed. The development plan may then call for further information to be exchanged, and approval sought, at other stages throughout the development period. Various changes may occur to the specification as a result of a change of need or through difficulties in meeting the original goals. Other changes may occur as objectives and needs are clarified by other development activities. Details then need to be renegotiated to ensure that the customer will still receive a product that meets the currently agreed specification. Each change results in an exchange of communications between customer and factory that has some effect upon the design or manufacturing activities being undertaken.

SUBCONTRACT-RELATED INFORMATION

Further cycles of information exchange take place between the factory and its suppliers or subcontractors. These may range from the raising of an order through to the development of a subcontracting specification which may itself be the subject of negotiations like those which originally took place between the factory and its customer.

FACTORY OUTPUT

Finally, in this simplistic overview, comes the factory's output. The declared output can range from paper clips to aeroplanes. Less obvious (and probably more difficult to pin down because of political considerations within the company) is the output that takes the form of wastage. The material sold off as scrap may be recorded, but it can be difficult to ascertain the proportions of swarf and damaged or irrecoverable items. There will of course always be losses that occur due to faulty paperwork, poor process control or because items just go missing for unexplained reasons!

INTERNAL FLOWS

Establishing this information flow across the company leads to a clear understanding of what activities are actually taking place, the order in which they occur and how they may be interrupted by interaction with the customer. Such a study identifies the major processes being performed and the degree to which they are dependent upon external resources and flows of internal information. In order to obtain an understanding of the detailed design and manufacturing activities and hence to establish the company's design needs (see Module 7.2), it is necessary progressively to subdivide the company as the study proceeds. The single black box with which we began is split up into smaller black boxes, representing departments, which are then further sub-divided, until we reach the level of individual activities.

DOWNSTREAM MANUFACTURING INFORMATION

The design process progressively creates a clearer and clearer definition and understanding of the proposed product. At the same time it also moves progressively towards a more and more detailed description of the manufacturing processes to be employed. Each feature of a component description should be developed with an appreciation of how it is to be produced. Every detail of the design must have a purpose (why else would it be included in the design scheme?), and all these details are therefore of use somewhere further downstream. To recognize this is simply to recognize the need to record all design information in an orderly manner so that everyone further downstream can have access to those bits which are relevant to their task.

For example, a blend radius may be calculated to provide the function of reducing the local stress value. But the size and form chosen for that radius also contribute information generated for the express purpose of instructing the pattern maker how to include this property in the proposed casting. A direct flow of information is thus required between designer and pattern maker which is currently being provided by the details on the drawing. Other drawn features and instructions have the express purpose of conveying information to other sectors of the organization. These include: materials details for purchasing, geometry and finishes for manufacture, inspection, information for assembly, and even instructions for dispatch.

All of these interactions result in information flowing around the company, passing from one black box to another. The identification of these flows allows the nature and scope of the real design process to be established. All too often plans for the introduction of CAD/CAM are based upon assumptions about what takes place within a company rather than upon what actually happens. The 'unofficial', but vital, channels of communication are ignored. It may, for instance, be assumed that all information is contained in the paperwork, even though much information is in fact transferred by means of accepted practices, informal meetings between staff from different areas or a general awareness of events occurring within the organization. Only if the totality of the information flows in all their forms is determined will a proper basis be developed for understanding the company's real design needs.

MODULE 7.2 ESTABLISHING DESIGN NEEDS

When attempting to define the design requirements of a company it is often far easier to see a whole range of things that could be done than to decide what is absolutely essential to the business's success. This lack of clarity is a result of the complex interactions between the various processes and their interdependency which makes it difficult to assess any activity in isolation. In order to distinguish the requirements which are essential from the activities which are peripheral or unnecessary the first step is to identify all the recognizable design-related work that is currently being performed within the company. This can be done by means of an information flow study (see Module 7.1). The various different activities and their interrelationships can then be investigated to establish how they fit together to form the design process as a whole (see Module 1.2).

THE DRAFTING PROCESS

Since the drafting process forms the 'pivot' between conceptual design and manufacture it is perhaps the best place to begin such an investigation. It is here that the general description of the product is transformed into an exact definition of a single manifestation of the original ideas. The final product is derived from a combination of the requirements contained in the design specification together with geometric information on relating parts, company standards, accepted good practices, advice from colleagues and the personal experience of the individual designers. The design is developed by a process of devising schemes that appear to meet the specified needs, testing these against the complete set of requirements and then modifying them until all the goals have been met. It is the designer's skill in coping with this array of requirements, constraints and possible solutions that is crucial to the achievement of a successful design.

By observing and recording the drafting activity it is possible to establish the level and quality of the information that the draftsman needs (and to determine how far his needs are currently being met) at various points throughout the development of the drawn scheme.

The process may, for example, commence with only an outline brief, in which case the draftsman needs to collect quantities of commercial or technical information before a scheme can be developed. Alternatively, the draftsman may be issued with drawings and schemes showing relating components and the assembly space within which the new design must fit, so that when his activity commences he already has an understanding of the geometric relationships described and must go on to deduce the additional functional relationships that need to be developed in order to produce the desired new scheme or item.

Both the type and the level of information presented to, or collected by, the draftsman-designer varies, depending upon the nature of the product and how far the project concerned has already been developed. If, for example, the item under consideration is one of the first parts of the overall scheme to be developed, then the draftsman requires a greater 'learning' time in order to gather the information and appreciate all the intricate details of the design. But as more items are designed, there is less need to learn and a greater need to relate the items under design to those already developed. The emphasis thus changes from interpreting technical requirements to ensuring the integrity of the geometric components designed to meet the functional needs.

STANDARDS AND CODES OF PRACTICE

The draftsman-designer has access to, and is in turn constrained by, the various company and industry standards. It is therefore necessary to establish to what degree he is aware of these standards and codes of practice. When consulted, are they seen as being supportive — that is a source of advice and alternative ideas? Or, on the other hand, are they perceived simply as irksome regulations and restrictions which must not be violated? If such practices are to be preserved within an integrated CAD/CAM environment, it is necessary to understand exactly how they have been used within the manual design processes.

SOCIAL INTERACTION AND EXPERIENCE

Whilst it is relatively easy to establish what information is used or consulted in the course of the drafting process it is very much more difficult to discover the amount of help, criticism and advice being provided by colleagues. In a well run department this may be substantial, though of course it varies according to the experience and authority of the individual draftsman. In the end every designer relies upon his own experience in order to select the most appropriate course of action when little real information exists or when conflicts arise.

Often, due to the limited availability of time and resources, one scheme must be chosen at an early stage and pursued to the exclusion of all other possibilities. When this is the case it is necessary to establish what information was available when the choice was made, whether better information could have been obtained and to what extent the full implications of that choice and the effect it would have upon subsequent activities or decisions were considered.

ASSESSING INDIVIDUAL PROCESSES

In this way each stage of the design process can be analysed by investigating the information flowing in, the subprocesses performed and the output generated. Only by assessing the output is it possible to ascertain whether a particular activity has been economic and effective. Often it is found that effort has been put into activities that were too complex, inappropriate or even irrelevant to downstream operations. As each part of the overall process is examined it is worth establishing how its output is used in the remaining design and production phases. This may reveal ways in which it could be changed so as to improve the communication of design data or to make the data communicated more relevant.

ESTABLISHING DESIGN NEEDS

Once the investigation is complete it should be possible to draw up a list of design needs relating to the various activities of the company. This list then forms the basis on which an assessment of the company's CAD or CAD/CAM requirements can be made. If, for example, vendors are to be invited to tender it is essential to indicate the extent to which it is necessary to produce schematic layouts, site plans and preliminary artwork. The advantages to be obtained by producing realistic, pictorial images (even in colour) can be assessed. The need for analysis tools, such as finite elements, and the extent to which they are likely to be used, can also be established. The complexity of the products involved determines whether the design process would benefit from the use of a full three-dimensional solid modelling system or whether one producing simple two-dimensional interpretations would suffice. The degree to which the final form is composed of complicated free-form surfaces indicates the type of geometric construction that is needed.

The nature of the manufacturing and assembly operations being undertaken dictates the form in which data should be provided to the drawing/design office. If there is a major commitment to subcontract work or product certification, formal drawings are needed which may have to match any one of a number of international standards. If, on the other hand, information is required only for internal usage then the individual needs of downstream operations must be considered and addressed directly. There may, for instance, be a requirement to produce numerical control data directly from the design model; or geometric data may have to be produced in order to facilitate other operations such as inspection, assembly, packaging or the provision of installation instructions. It may be advantageous to treat the generation of this information as an integral part of the design process rather than an additional operation occurring downstream as at present.

Time spent in such a study is not wasted, even if the company already has a CAD/CAM system or has no intentions of purchasing one. The study provides the investigators with a complete understanding of how the company actually conducts its design activity. It is only rarely that no surprises emerge. People are usually amazed at the complexities that are revealed and are often embarrassed to find that senseless and wasteful activities have been allowed to continue for so long undetected. The identification of this waste, and its rectification, can be quite sufficient justification for the work involved in conducting such a study.

MODULE 7.3 IDENTIFYING BENEFITS

DRAWING PRODUCTIVITY

In the past it has been the normal practice to base the justification of a CAD/CAM system upon a calculation of the benefits that can be achieved by greater drawing productivity. But though such gains are easy to measure, this approach is based on the twin assumptions that a greater output of drawings is beneficial in itself and that unneeded draftsmen can in some way be 'displaced'. In other words, a system is justified by balancing the 'value' of the additional drawings that can be produced with its help against the cost of the additional staff that would be needed to produce those drawings manually; alternatively, if no expansion in drafting activity is envisaged, the cost of the system is set against the savings that can be made by reducing the existing staff. On an accounting basis such a case is easy to construct and its simple logic can be accepted by everyone. The payback time is calculated by dividing the annual saving in salary costs into the total cost of the system, and the calculation can be made to appear 'more realistic' by making suitable deductions for maintenance, services and depreciation.

Unfortunately, this kind of approach gives a totally false impression of the benefits that can be expected from the use of a CAD/CAM facility. No matter how the calculations are dressed up, they remain bogus and represent a vastly over-simplified view of the effects of installing a system. The major misconception is the idea that CAD/CAM can, somehow, replace people; that it can perform all the day-to-day tasks of a draftsman-designer. This is not the case. The system is a tool and, like all other tools, has to be driven or directed by an external intelligence.

INITIAL SAVINGS

Like all technological developments, the installation of CAD produces an initial saving as mundane and boring tasks are taken over and automated. However, savings of this kind rapidly level off as the more complex inputs, controls and decisions that have to be provided are far more difficult to define and handle automatically.

MINIMUM STAFFING LEVELS

The minimum number of staff required to maintain a company's design activity is often greater than might be suggested by simplistic calculations of system productivity. For instance, if the figures suggest that only a fraction of a person is required to conduct a particular design activity, is it wise (or practical) to entrust this, perhaps highly significant, aspect of the business to a 'part timer'? In order to ensure continuity of design in the event of illness or resignation, it is often preferrable to maintain a minimum of two members of staff operating in each area of the company's business. This also improves the design process by providing the opportunity for interaction and creative support within the design team.

In many companies this minimum level of staffing may have already been reached. There may no longer be a large pool of well-trained draftsmen ready and able to produce the company's drawings.

Indeed, most companies have suffered badly from a shortage in design skills and qualified staff are difficult to find. This may in fact have been the stimulus that has led them to consider CAD/CAM in the first place. It would clearly be folly for such businesses to suppose that they can justify the cost of a system by reducing staff levels which are already at a minimum.

DOWNSTREAM JUSTIFICATION

If it is impractical to reduce staff and if no increase in drawing output can be predicted, how can a system be justified and its benefits demonstrated? The answer must be to consider its effect upon the downstream processes. Data gathered on the information flow and design needs of the company (Modules 7.1 and 7.2) demonstrate what information is really required at each decision stage. It should, therefore, be possible to establish how a CAD system might help to furnish that information, or to provide better information, and a justification can then be based upon the cost savings that will be achieved.

PRODUCT DEVELOPMENT CYCLE

In order to estimate where and how these savings will arise it is necessary to look at the overall product development cycle time and see where the major delays or process iterations occur. What are the principal unknowns that cause problems or necessitate redesigns? How could they be contained or eliminated? If a component is tracked through from conceptual design to ultimate manufacture, it is found that it is being worked upon (either intellectually or physically) for a very small proportion of that cycle time. For most of the time, the design is 'idle — awaiting action'. The drafting time rarely takes up more than 10% of the cycle and so its reduction, even by a half, is of insignificant effect in itself. However, using the time saved in drawing to identify and eliminate potential problems at an early stage has been shown, in some cases, to save up to 40% of the complete development time. This sort of economy is clearly worth investigating.

Major improvements in development cycle time can be achieved in many different ways, depending upon the type of industry and the constraints imposed upon the design process (see Modules 1.1 and 1.2). Where they can be found depends on how the company secures and maintains its customer base. A number of reported improvements and their benefits, both direct and indirect, are now described to illustrate some of the possibilities that may repay investigation.

ACCURATE GEOMETRIC DESCRIPTION

A company designing and manufacturing process equipment for installation within customer's factories considered the benefits CAD/CAM might bring to their business. In this case CAM would be of little help since very few of the component parts were produced in-house. The subcontracted items ranged from batches of small turned parts to the structural steelwork for the plant (complete with gantries and catwalks). All of the large structural items were shipped directly to the site whilst the smaller components were delivered to the factory for assembly into the special purpose machinery. These machines were then individually built and tested before being transported to the customer's site. A considerable time, perhaps many months, would then be required to assemble the gigantic three-dimensional jigsaw puzzle on site and commission it to the customer's satisfaction.

An investigation of this phase revealed two important facts. Firstly, the customer was not really sure what he was finally getting. Secondly, due to the complexity of the design and the inherent uncertainties in the way it had been developed, everyone tended to err in a positive direction when they were unsure of an actual size (on the assumption that it would be easier to remove the surplus on site rather than to remanufacture a part that was too small). At bad times, a site could be littered with people adjusting parts by burning bits off. This spectacle added to the customer's worries. It was also seen to be a very expensive activity for which the customer did not expect to pay.

The cost of investing in a CAD system was therefore compared, not with the economies that might be expected at the drafting stage, but with the savings that could be achieved through utilising its ability to model the complete plant in three dimensions within a revised design process. Initially, the customer's site would be described and all reference features included. The main details of the process equipment under design could then be added. This site model could then be verified and approved by the customer, who would be fully aware of what was being proposed and was responsible for checking that the site had been described correctly.

All major assemblies were then detailed from these reference features, with the position of each subassembly or individual machine being related to reference points set into the major assemblies. By defining and maintaining known spaces for the machines about these reference points it would be possible to ensure the correct assembly of the complete plant by inspecting and verifying only the reference data between items, no matter how large or small.

The development of these design 'control' processes would, it was predicted, result in a saving of up to 75% of the erecting and commissioning phase times. Due to the high primary cost of such process equipment and the relative simplicity of the modelling requirements, the payback calculations indicated that the system would break even on the first contract and show major saving from then on.

ACCURATE JIGS AND FIXTURES

The savings often quoted for the aircraft industry can be impressive, both in the area of design and in the manufacturing process. Many design and analysis activities can now be undertaken that would have been quite inconceivable without the tools provided by CAD/CAM and its interface to large analysis facilities.

Here, however, we shall concentrate on the downstream manufacturing gains; these may be less spectacular, but they are applicable to a wide variety of industries. They take a variety of forms. When using normal design and manufacture procedures it is not unusual to find that a stack-up of errors in one direction makes it difficult to align a sub-section of the airframe with its jig and mating parts. Similarly the build-up along the fuselage length may mean that floor panels have to be individually fitted. Jigs and fixtures are often unavailable when a prototype is built, because it is impractical to determine what will be needed until the problems of manufacture or assembly have been confronted. The needs of the assembly crew, in the form of towers and gantries, to allow appropriate access to parts of the body, wings and tailplane are often considered only when the crew is attempting to create the first assembly by using temporary scaffolding platforms.

The use of CAD/CAM improves component definition and eliminates the build-up of errors thus ensuring that all parts will fit; it also allows the jigs and fixtures to be developed and produced in time for the prototype production. This speeds up the manufacturing phase and significantly reduces the product development cycle.

CONSIDERING ALTERNATIVE DESIGNS

As an aircraft is expensive both to produce and to run, there is little point in producing it even, say, a month before the airline wishes to take delivery. It will sit around unused, costing either the manufacturer or the buyer money. The advantage of a CAD/CAM system is that the design could, in this case, be started one month later! Whilst this may take a considerable amount of courage to do, there are gains to be made beyond simply avoiding having an expensive aeroplane lying idle prior to delivery. The initial delay in committing to a particular design process can be used to advantage in considering alternative design strategies and markets. Can additional customers be found if slight changes are made? Can simplifications be made that will not compromise the original scheme? Can alternative materials and processes be incorporated to lighten the design or reduce its manufacturing costs?

Whilst the advantages so gained are difficult to quantify in a general case, they can be very significant in particular instances. Reductions of 10% in manufacturing costs and doubling of potential sales have been quoted. Whatever this 'front end' benefit amounts to, it is certainly additional to the savings that can be achieved by reducing the complete product development time.

REDUCTION IN DELAYS

Justifications for CAD/CAM systems are often based upon the savings that can be made by reducing expensive delays, and it often turns out that there are additional advantages which no one had foreseen. We consider an actual example which concerns the process of creating masters for an optical line-following mill. In order to achieve the necessary accuracy, this is usually done by creating master artworks at ten times full size which are then reduced photographically. The procedure involves the detailing of a large-scale master, the creation of the artwork which is then sent away for reduction and, finally, the testing of the resultant copy. In the case under consideration these activities could extend the process cycle time by up to six weeks.

The use of CAD allowed the line following master to be produced directly from the system's plotter at the correct size (and with greater accuracy than was achieved by the photo-reduction process). Such plots could be produced in hours as compared to weeks by the old process.

In practice, however, the main advantage came, not from the time saved, but from the greater usage that was subsequently made of that machine tool. Practical problems arose when the reading head was required to follow a complex or sharply changing curve; there was a risk that the head would be unable to follow the path correctly and that errors would be created in the machined part. When changing the path involved a six-week delay, designers either took no risks in their designs or they ensured that they were not manufactured on the line-following mill (resulting in more expensive products).

Once CAD was introduced, faults could be corrected and changes made to the plots in a matter of minutes. There was no need to strip down a job whilst these changes were made (or even for anyone else to be aware that these had taken place). The result was that the number of jobs coming to this machine steadily increased, improving both its productivity and that of other machines.

OTHER AREAS OF BENEFIT

The benefits of a system may be assessed in many ways, but are always very dependent upon the particular case being studied. It is thus not possible to assume that gains achieved in one context will necessarily be repeated in another. It is, however, worth briefly indicating other areas in which significant advantages that may contribute to the justification of a CAD/CAM system have been recorded.

The generation of detailed plans and specifications for quotes can, if the design is based on an assembly of 'standard' company modules, be successfully carried out on a CAD system. A reduction in man-hours to less than a quarter, an improvement in quality and a tenfold increase in 'capture rate' have been identified in particular cases.

A manufacturer of selected parametric parts has indicated that savings of over 50% are achievable by directly linking the design database to machine tools. The overall benefits to any other such factory must, however, be assessed by investigating the proportion of standard or common parts that could be employed in a changed company design process.

Finally very large gains have been established in the area of inspection, particularly in automatic coordinate measurement. Here no added value arises, since the process only consumes time and can at best indicate that errors have already occurred. But rapid in-process inspection can identify and rectify problems before too much scrap has been generated. The reduction in costly inspection time alone can produce a payback upon a system well within two years.

INDIVIDUAL JUSTIFICATION

The assessment of the benefits of CAD/CAM systems is seen to be not simply a matter of applying accepted productivity calculations in a straightforward accounting exercise. All systems need to be justified on an individual basis by studying the product design and development activity together with the information flow throughout the company. When all this has been done a proposed payback period can be estimated. Whilst all such exercises are subject to a degree of uncertainty, the greater the depth of the study the less chance there is that the decision to invest in CAD/CAM will be made on false assumptions or in the expectation of benefits that are, in practice, unobtainable.

MODULE 7.4 TRAINING FOR OPERATION

USE OF VENDOR TRAINING

In order to achieve the best possible return on an investment in CAD equipment it is necessary to provide a course of suitable training for the staff. In many cases a company may consider it sufficient to send a limited number of designers, draftsmen or managers away on the vendor's training courses and then to use them to train everyone else. Sometimes those chosen for the initial training are selected because of political considerations (rather than technical need) and their numbers are determined simply on the grounds that so many free training places are provided as a part of the vendor's package. Having paid a relatively large sum for their system, many companies are surprisingly reluctant to spend a comparatively small amount in order to ensure that their staff can use it effectively.

REASONS GIVEN FOR LIMITING TRAINING

Various reasons have been put forward to justify this position. These may range from straight financial arguments to, for example, the structure of a company's training policy. It may be, for instance, that the initial cost of the system is met out of central funds while the cost of training courses has to be met out of limited departmental funds. Other companies operate a policy of minimal CAD/CAM training for operators on the grounds that the higher the level of skill obtained the quicker they will leave for a better paid job. This may, admittedly, be a short-term problem, but to deal with it in this way is like cutting off one's nose to spite one's face. The required productivity and design improvements cannot be achieved by poorly trained operators no matter how long they remain at the company.

SKILLS REQUIRED

Reliance upon vendors' introductory training courses results in all staff being competent at a uniform, basic level. But skills then need to be developed to a level appropriate to that at which the operator is required to work. It is thus necessary to consider the range of skills to be developed in the individual and to construct a training programme accordingly. These skills may be split into three basic types.

CAD/CAM AWARENESS

A large number of people in the organization need to know how to relate to the new system, but not all need be competent operators. Initially, however, everyone must be put through some kind of awareness course. The broadest such course, and the least job specific, should be designed to allay fear of the new technology. It should be used to demonstrate how the system is used, how the operator interacts with it, and how the information can be retrieved and used elsewhere within the company. The bulk of the course should consist of 'non-expert' demonstrations, discussion sessions and hands-on experience.

Training for Simple Interaction

A smaller group needs to be trained to relate to the system, and to collect information required for their jobs, but still not to become full operators. Once these staff have been provided with an overview of the system and its processes, areas of specific interest to individuals can be dealt with in greater detail. As processes change or software is revised it is necessary to provide retraining courses for these users. The updating courses should also be used to provide feedback on how effective the system is, how it could be improved or, indeed, whether staff are actually bothering to use it (when change occurs it is sometimes easier to stay with the old methods rather than put effort into learning the new).

Full Operator Training

The approach taken to the training of designers and draftsmen depends upon the specific local definition of these titles and the associated roles. In some companies, the draftsmen have simply taken over the role of operators. The designers, as in the past, develop rough schemes and perform the detailed analysis that is necessary. The draftsman/operator is then required to enter that data into the system to produce the model much as he might be deputed to construct a drawing on the drawing board. The designer need not then be trained to the level of full operator. He must, however, be aware of the techniques and processes involved in the generation of CAD models in order that he can exploit them to the full.

When a company operates with its design and drafting staff taking equal responsibilities on the system then all designers must be trained to the same level as the draftsmen. Although at first it may seem that conflicts will arise as to who should do what upon the terminal, these problems disappear as roles and responsibilities are worked out. The designer should be responsible for the generation and development of the design scheme. The draftsman's role is then, as with the drawing board, one of detailing and specifying the chosen arrangement to company and national standards, in a manufacturable form. The training provided for each group should be tailored to reflect what is required of them.

System Management

Beyond these three basic training needs lie two others of a specialist nature that must be addressed. The first is that of the system manager and his team. They are the first point of enquiry when problems arise. Typically, their tasks range from explaining how to adjust the workstation chair right through to unravelling a corrupted database. In a large management team, it may be possible and practical for individuals to specialize in various types of user problems. But whether the team is large or small, the operator with a problem expects it to contain someone who can answer his problem. There is thus a need for these staff to be provided with a very high level of training. It should cover all aspects of the system from daily maintenance and preventative procedures through to usage of advanced techniques.

SYSTEMS DEVELOPMENT

The second of these specialist requirements is for systems development. The very mention of 'systems development' strikes fear into most companies. Their attitude is that they have bought a system that is supposed to work, so why should they be responsible for its development? It is also assumed that development is a disruptive activity that will produce a continuous stream of up-grades. To think this way is to misunderstand the role of the systems development staff. It is not to build a completely new and better system but to tailor the existing one more closely to the company's specific needs. They should be developing the company's database, ensuring that efficient means are provided for accessing it to retrieve similar objects and common parts, and for the identification of previous solutions. They may identify specific aspects of the company's design process that could be eased by the creation of a suite of special programs. It is they who provide tailored menu programs for staff and develop procedures to control or restrict access to specified levels of information.

These systems developers may be treated as a part of the general management team, but they do require skills in programming and operating systems beyond the normal level. Specialist training must be provided.

ON-GOING TRAINING

Far too often, training is considered to be a once-and-for-all operation; if the man is trained, he is assumed to be trained for life. This is not the case, for many obvious reasons. Firstly, the system does not remain unchanged forever. There will be regular up-grades and enhancements provided by the vendor. The company may wish to acquire additional software to address new markets or skills. All such developments necessitate some degree of training if they are to be used effectively.

Even if the requirements arising from system changes are set aside it is still desirable to provide a continuous training programme for all staff. Such a programme should be provided and monitored by the management group for each member of staff. It should be recognized that no one retains and uses all the information provided in one training session. Some skills, once learned, are quickly applied in performing real work and so are reinforced. But many others, not immediately put into practice, are forgotten. Periodic refresher courses should be provided, reviewing the skills learnt and identifying the areas in which additional skill development would be beneficial. An underlying goal should be to raise progressively the level of skill throughout the organization in order to realize the maximum benefit of the CAD/CAM facility throughout the whole range of the company's business activities

MODULE 7.5 WORKING ENVIRONMENT

THE PROBLEMS

All too often little effort is put into arranging the workplace once the CAD/CAM system has arrived. The attitude is that everyone should be getting on and using it (to achieve the benefits and payback as quickly as possible) rather than considering the advantages of planning for its best use. Perhaps most people are so overawed by the cost and complexity of the system that they find it difficult to understand how they can or should organize it. It is seen as an object to be 'approached on its own terms'.

The result is that the terminals are badly laid out from an ergonomic point of view. This is often apparent to an outsider when the terminal area is first seen. It may have been constructed for the computer rather than for the human operators. Some users complain that computer systems now have the environment that the workers have been seeking for years.

AIR CONDITIONING AND HEATING

Air conditioning is sometimes seen as a necessity for the computer but not for the staff. Most modern workstation configurations do not, however, require a complete air-conditioned environment. It is more important to provide a relatively stable temperature and humidity range, eliminating the extremes of winter and summer. But, given that the system will operate effectively within a temperature range greater than that which would be tolerated by most operators, it is sensible to provide a facility in which the temperature and humidity are comfortable for the users. This both improves the reliability of the system and the amount and quality of the users' work. The productivity of the system is after all limited in the main by the rate at which the users can interact with the developing designs.

LIGHTING

A question usually associated with the heating problems is that of the correct level of lighting. The first things to be considered is the position of the windows in the workstation area. Whilst windows reduce claustrophobic feelings, they do provide large sources of uncontrolled light. It is therefore usually necessary to provide some kind of venetian blinds to diffuse the natural light. Terminals should then be positioned sideways onto the windows whenever possible; if this is not done the operators will either be looking into light coming from behind the workstations or having it reflected off the screens from window behind them. A sideways position is the best compromise but even this may be difficult if windows are present in more than one wall. In this case, the level of natural light needs to be checked and decisions made as to whether opaque screens or curtains should be used. Often the arrangement cannot be finally decided until the actual workstations are installed as their brightness and clarity affect the acceptable level of illumination in the work area.

Frequently, the light level of the terminal screen is set too low. This has been recommended in the past to preserve the life and quality of the display, but the technology used in screens has now reached a point at which this is no longer necessary. As a consequence, the ambient light level need not be exceptionally low. The aim should be to strike a balance between screen brightness and light levels at a point where the operator's eye fatigue is minimized. Constantly switching from looking at a bright screen to studying dimly lit notes can be very tiring. The more uniform the lighting level throughout the working area the less stressful it is for the user.

INDIVIDUAL WORK AREAS

Whilst the operator may wish to be able to communicate with his fellow operators he may also not wish to be disturbed by them. It is possible, in an open area, to provide some degree of privacy by a suitable arrangement of lighting. This can be used to provide islands of illumination around each workstation within a lower background level. It is however usually necessary to provide additional noise insulation in the form of screens which can also be used as mounting boards for the (usually large) collection of information that the designer takes to the terminal. Care should be taken to ensure that each workstation does not come to resemble a confessional booth which is open to everyone but belongs to no one. Some companies, in an attempt to overcome this tendency and to generate a degree of commitment to the workstation operated by an individual or group, have encouraged them to be 'customised'. This has resulted in a range of workstation styles ranging from art deco to tropical botanical!

SEATING

Once at the workstation the operator needs to be sitting comfortably.

On a day-to-day basis, the arrangement between man and workstation is rarely considered. A number of ergonomic terminal arrangements have been marketed. They have not been widely used, not because of non-availability or cost, but through a lack of demand from the users. Most operators have not been educated in the advantages to be gained by selecting the correct ergonomic posture at the terminal. The most comfortable position is achieved when the trunk is held slightly back from vertical with the thighs well supported. The upper arm should hang nearly vertical whilst the forearm and hands rest horizontally upon the keyboard. Finally the head should tilt down slightly to view the screen at a comfortable distance of approximately 0.8 metres. To reduce reflections the line of sight should be normal to the screen. These various requirements can only be met if the height, attitude and relationship between seating, desk, keyboard (or tablet) and screen are adjustable for each operator. As it is rarely possible to vary the desk height, it may be necessary to provide adjustable foot stools.

BENEFITS OF GOOD ERGONOMICS

Careful evaluation of the needs of individuals and suitable adjustment of the workstation arrangement has been shown to reduce operator fatigue considerably. The number of minor complaints also drops when the operator is happy at his workplace. It is, however, not easy, nor should it be necessary, to put a cost advantage upon the comfort of the user. One could be silly and attempt to equate the costs against the hospitalization fees or days lost through backache or eye strain. The truth is rather that the success of the installation is completely dependent upon the amount of creative work that the operator can perform on the system; any improvements that will lead to a greater or more continuous period of operator creativity must be seriously considered.

MODULE 7.6 DESIGN AND SOCIAL STRUCTURE

NEED FOR PLANNING AN INSTALLATION

It may seem to be an obvious thing to say, but no system should be installed without there being a very well-developed plan as to how it is to be used. Too often, after the considerable effort put into the evaluation and justification exercise, the period spent waiting for a boardroom decision and the subsequent delivery time may seem to be an anticlimax. But this unavoidable interval provides valuable time that should be used to consider the business requirements and to formulate a plan for the introduction of the CAD/CAM system into the company. The planning should take into account the design requirements and the social structures that currently exist and those that will result from the introduction of the new technology. Rarely is it possible or desirable to install a factory-wide CAD/CAM system in one go. This is not only expensive but also very disruptive to the organization. Considerable efforts (and some luck) are required if mistakes are not to be made on a grand scale. As so many changes, planned and unforeseen, can arise from the introduction of the system, it is unlikely that no change has to be made to the original plan during its implementation. However this is not an excuse for not producing one; it is better seen as a justification for a phased introduction. An overview must be created for the complete system in broad terms in order that each department or group is aware of their position and role in the new structure. Detailed plans need then to be drawn up for each stage of the implementation.

PROCESSES TO BE PERFORMED

A number of major factors influence the effectiveness of a CAD/CAM facility and therefore need to be taken into account in developing a structured plan. The first thing to consider are the activities to be carried out on the system and their effects upon the company's design process. Those activities that are required for the successful development of the product should have been established (Module 7.1) and the ways that information is communicated between them identified (Module 7.2). This information, together with an appreciation of how the general design process fits in with the product needs (Module 1.2), provides a complete understanding of the design process being used in any particular instance.

This is different for different companies, depending upon their size, the industrial sector in which they operate and the constraints on their products. These studies do, however, permit the 'heart' of the process to be established. This may also be looked upon as the pivot about which all activities revolve. The processes may centre upon a complex and detailed analysis of the product performance, as is the case with finite elements. Alternatively there may be many separate activities, including styling, tool design, and value analysis, that all comment on, or contribute to, the selection of the finally chosen design. In this case the pivotal process is the development of a detailed geometric model.

PIVOTAL ACTIVITIES

It is these pivotal activities that should be first addressed in devising the implementation plan for the CAD/CAM facility. No decision can be made on the peripheral processes without knowing how they are to impinge upon the central one. The major benefit of the system will only be realized when the outputs of this central, pivotal process become available, firstly, in all related evaluation and development activities and, secondly, to all dependent downstream processes. It is thus crucial to decide at the outset how such information is to be handled and made available to those who need it. The result of this decision influences the way in which, and the order in which, all other activities should be brought within the scope of the system.

How the facility is to be developed around this pivotal data depends to some extent upon the industrial sector that the business is in. Some areas are strictly regulated by national and international regulations and codes of practice. Some require complex or time-consuming analysis activities to be performed. Some depend to a large extent upon standard parts provided by a well-developed sub-component industry. Software houses and vendors have developed suitable support programs for companies subject to each of these sets of considerations.

If, for example, the company is intending to construct a CAD/CAM facility for the development of electrical instruments based on printed circuit board technology, then a wide range of software ranging from circuit analysis to robot insertion of components is currently available. The implementation plan is based upon selecting that which is most suitable and deciding, from the design process studies, how and when it should be introduced.

A company that is intending to use their facility for the development of complex and specialised mechanical engineering products, on the other hand, may well find that only general computer aids exist. The integration into one computer package of, for example, mechanical design and optics to produce camera lenses in the form required may have to be undertaken by the company itself or contracted to consultants. The implementation plan in this case would require a commitment to, and an understanding of, the developments that need to be undertaken during the introductory phase.

INTERACTING FACTORS

The way these design needs are satisfied is thus dependent upon the available technology and the demands of the product. How the new techniques are introduced depends upon the receptiveness of the staff. This, in its turn, often depends upon the size, age and location of the company and upon the level of technology it currently uses. As all of these factors interact and depend very much upon local conditions, it is difficult to say for certain where the greatest difficulties are likely to be encountered. The response of the work force is partially governed by the level of trust that exists within the organization. Trust is fragile and can take years to build but only a moment to destroy. The following comments are thus generalizations that need to be carefully tempered by knowledge of any particular local situation.

Company Size

Communications within a small company are generally more informal and more direct. Problems are handled by the parties involved at the level on which they occur (there may be no other 'level' within the company!). The larger the company the more likely it is to have a negotiation structure that removes any major problem from the individuals concerned up to a level at which principles and procedures are discussed and agreed. Within a major, international, high technology company, times of working at a CAD/CAM workstation may have to be formally agreed and adhered to. Small companies can take a more flexible approach. One such company is known to have increased the utilization by allowing some staff who were keen fishing or gardening enthusiasts to pursue their hobbies during part of the day and work either early or late sessions. Another company, which works on a project team basis, makes the system available 24 hours a day so that the teams can choose for themselves when they wish to work.

Existing Level of Technology

Often new, small companies are already in the area of high technology and their staff are used to handling advanced equipment and novel ideas. For them, CAD/CAM is just another innovation among many, and can be easily assimilated and cause no problem or reaction. It is in the larger, low technology companies that problems can arise. Such companies may operate their manufacturing plant without any computational facilities (their only experience of computers may be when the accounts department gets their wages wrong!). The lack of understanding of what is being proposed, how it will affect them and how they will cope, results in a threatening situation and may provoke a hostile reaction. In such businesses the implementation plan must proceed slowly and carefully. Each phase of the introduction must commence with appropriate discussions with the staff concerned and be developed with easily defined and achievable goals. The design processes and social structures must, initially, be built around the existing departmental arrangements and slowly modified as further aspects of the developing process are included.

Departmental Structure

The most obvious result of the introduction of CAD/CAM techniques is often the integration of various activities previously carried out in different departments. It demands a greater allegiance to the project or products than to departmental groupings of skills or equipment. There is therefore a natural tendency for interdepartmental groups to be formed. If this results in improved product and performance then this should be encouraged at the expense of the formal departmental structure. Care should be taken to ensure that the traditional processes are not abandoned without providing alternative regulating and monitoring procedures.

The selection of a design and social structure appropriate to the needs of the company can be seen to be a complex and emotive activity. It can range from a large and formal structure through to an informal process controlled solely by project needs. Whilst the company and product constraints dictate to some extent what can be accomplished, the evidence is that greater success is likely to be achieved when companies are prepared to move away from the traditional departmental structures towards those based upon groups formed to handle the demands of projects or products.

MODULE 7.7 SYSTEM MANAGEMENT

The management of a CAD/CAM facility can be a thankless task. If it is done well, no one should know that it is happening and the only time people will be aware that someone is in charge and responsible is when it all goes wrong.

MONITORING USE OF CORRECT PROCEDURES

As most operators are under pressure to produce work against deadlines, it is not surprising to find that the rules get broken or forgotten. Such pressure may, for example, cause someone to change a print by hand rather than spending time modifying the model on the CAD system. But if this is not subsequently done, there will be two manufacturing standards: the marked-up drawing and the computer model. Again, it it tempting, when in a hurry to meet a deadline, to file an item under any convenient name rather than going through the company naming and numbering procedures. But if this becomes a regular practice drawing files become lost in a sea of *ad hoc* names and are difficult to trace.

The system manager must try to ensure that the correct company practices and procedures are adhered to, no matter how tiresome this may be to people working under pressure or against the clock.

USING CAD AT THE APPROPRIATE TIME

Besides his obvious monitoring role, it is necessary that the system manager should be responsible for a whole range of design and 'domestic' activities. The first, and perhaps most important, job is to ensure that the design work arrives at the system at an appropriate time. This is not just when there happens to be a terminal free, but at the first moment that the geometric form is being specified. Opportunities are lost through designs arriving too late. In one real-life instance designers were working out the unfolded shapes of sheet metal shutes using cardboard models and then asking the CAD operators to draw them on the system. The software being used was in fact a full three-dimensional modeller which had the ability to unfold automatically sheet metal forms! Had the system manager been allowed to interact directly with the designers and handle the complete design operation, much time and cost would have been saved.

There are, of course, also examples where the CAD facility becomes involved at too early a stage, often because it is assumed that, somehow, it is able to create data and generate a specification by itself. Most detailed designs are dependent upon knowledge about mating components and internal properties and function. This information must be already available for the design to be successful. If the design work to be undertaken is discussed openly at an early stage then the right level of knowledge can be brought to the system in order that the CAD modelling can start at the earliest possible time.

PRIORITIES

Once the correct time for starting to involve the CAD/CAM facility has been agreed, it is then the manager's responsibility to discuss and agree the priorities to be attached to current and new work being undertaken upon the system. Priorities should not be solely controlled by the ability of a person to get his name down first on a booking form. Such forms should be used to indicate the currently agreed terminal loading, not the priorities. These should be negotiated on a needs basis.

The booking system is all too often taken as part of the rules of conduct rather that as an indication of loading. Some organizations allow large blocks of time to be booked wastefully, whilst others throw off an operator after a specified time interval, irrespective of how long it may have taken him to retrieve the information or how close he is to completing the set task. The manager must thus play a sensitive role, balancing the conflicting needs of the users against the need to ensure that completed and successful products emerge on time.

MAINTAINING THE SYSTEM

The other 'domestic' activities that consume the manager's time relate to maintaining both the system and its engineering database in an efficient operating condition. These include carrying out and monitoring the routine maintenance procedures, upgrading the software, archiving current data and retrieval of previous items. Such activities are to a large extent dependent upon the size and configuration of the system being operated (Modules 2.1 to 2.5).

SYSTEM MANAGER - A KEY COMPANY ROLE

Ultimately the success of any CAD/CAM facility depends upon the system manager. It is he who develops and encourages successful usage. It is he who determines what procedures are developed to aid the company's design process. If he is to achieve his objectives he must be supported throughout the company. His actions must be seen as reflecting the desired activities of the whole organization and not just as expanding the 'empire' of the CAD/CAM group. It is therefore necessary that he has managerial status and is involved in the decision-making process. If this is not done, his status will be that of a machine minder and the system's full potential will never be realized.

EXERCISES

1. Identify a local engineering company and consider the flow of design and manufacturing information necessary for the development of its products. If a real company can be approached, so much the better. If one is not available consider the likely flow occurring in one of the following organizations:

a company sub-contracted to produce printed circuit boards to order

a company contracted to develop and install special purpose engine test equipment for a motor manufacturer

a company manufacturing tape recorders for the mass market

a company producing plastic bottle moulds for the cosmetic industry.

2. Using the companies listed in exercise 1, identify the areas in which each is likely to benefit from CAD/CAM. How should they reorganize themselves to take advantage of the possible gains.

3. Consider what information is required for a draftsman to produce a drawing of an engine connecting rod. You can assume that all other decisions have been made with regard to the engine and component designs. Identify initially the geometric association with other parts and possible modes of failure or clash. Then consider those features necessary to carry the loadings and the likely need for numerical calculations.

4. Now consider the downstream effects of your chosen connecting rod design in exercise 3. What features need to be described in order that it can be manufactured? How will these differ if the rod is to be machined from solid material rather than being cast or vice versa? If these manufacturing factors (and those of engine maintenance) were taken into account how could the original design be improved?

5.What kind of training programme is necessary to prepare the staff in a large company for the introduction of a CAD/CAM system? Assume that the company is one in heavy engineering that has a traditional departmental structure and no current experience of computers.

6.What role should be defined for the manager of a large (central) installation that is serving design, drawing office and manufacturing departments through (remote) terminals located in different departmental areas? Describe the range of responsibilities and indicate the status that is necessary for him to be able to carry out these tasks effectively. Suggest also the level of support necessary from the company management structure.

7. Select a real company (either one that is known to you or approachable locally) and through a brief visit and discussions establish the pivotal aspect of their design process. Based on this investigation, indicate, firstly, what CAD modelling techniques would be most appropriate to their product and then suggest how CAD/CAM could be introduced.

8. Indicate why the social structure to be found within the CAD/CAM facility of a major manufacturer of consumer products is likely to be different to that found in a small, high technology consultancy company specializing in the creation of advanced electronic systems.

9. Describe the major factors to be taken into account when laying out a new terminal area to be used by many operators. Indicate the considerations that have to be weighed in assessing lighting needs, noise levels and the arrangement of local work areas.

10. In order to achieve the most stress-free working conditions, what are the best ergonomic arrangements for operator, terminal and seating.

11. Indicate the various ways that a CAD/CAM facility can be tailored to suit the needs of a company. Choose a particular industrial company, such as one specializing in the manufacture of high-pressure boilers, and suggest how aspects of the design process could be handled by the generation of computer programs and also how the various company and international standards could be made available using the facility.

12. Consider a company designing, manufacturing and marketing a range of special purpose packaging equipment (such as would produce boxes of frozen peas). They manufacture the mechanisms in their own workshop whilst sending the large parts (such as the frames) out to subcontractors (who do not have CAD/CAM). These machines are shipped and installed all over the world. How should they set about organizing themselves to investigate what CAD/CAM will do for them?

SECTION 8
APPLICATIONS PROGRAMS

As we have seen in earlier sections CAD/CAM systems can be used to assist the designer at all stages of the design process (see Module 1.4). Most general-purpose systems are, however, designed primarily for use at the scheming stage (see Module 1.2). Indeed, the core of any system is its geometric database within which a full description of the product design is generated. For without the geometric modelling and manipulation capability it is not possible to perform any of the other design and manufacturing activities.

Although it is the database, containing the model description, that forms the heart of every CAD system, it is usually more convenient to think of a system as a database plus the graphical processes that are employed in the display and plotting of the constructed images. The basic CAD/CAM system can then be described as a computer within which geometric models can be created and displayed.

The form in which the model is held varies considerably, from simple two-dimensional views through to full three-dimensional solid geometry. The programs that manipulate these models to provide particular forms of output images could be considered as applications packages in their own right but they, too, are normally 'lumped in' with the basic system. The ability to create engineering drawings and pictorial images is considered as a primary requirement for any system and not as an additional item to be purchased separately.

The applications programs to be considered here are those software packages that interact with the geometric database, either to aid in its creation or to extract and use the information which it contains. A vast array of such programs have been written to address a whole range of problems that have been identified, relating to every stage of the design to manufacturing process in all sectors of industry. A glance at the long list of applications programs on offer from most CAD/CAM vendors reveals a bewildering variety, ranging from programs which handle simple equation manipulation for the design of such objects as truss frames for buildings, through to others which can perform analyses of thermal loading within electronic circuit designs.

Applications programs have been developed for many different industries, and the complexity of some applications, or the level of knowledge that is required about a particular industry before they are understandable, makes it difficult to discuss many of these specialized programs within the restricted format of a book such as this. The applications programs described in the following modules have therefore been chosen from those used in the mechanical or the electrical industries on the principle that the reader is more likely to have a broad understanding of these industrial activities, than of subjects like cartography and chemical plant design.

The application programs have been grouped into modules according to which stage of the overall design process they relate to. Many of the programs designed to carry out analysis are directed towards the resolution of over-constrained problems; programs that relate to downstream manufacturing activities, on the other hand, tend to focus more on under-constrained problems. The need to exercise control over the complete design-to-manufacturing process has led to an increasing number of programs being offered which are designed to aid in the planning, management and control of the various operations and activities. This development has, in turn, led to the introduction of programs based upon expert systems technology in order to help companies integrate all these aspects of their business into a single design process.

MODULE 8.1 ANALYSIS-CENTRED APPLICATIONS PROGRAMS

The place of analysis activity within the overall design process means that it interacts directly with both the conceptual and the scheming stages (see figure 1.8). In some circumstances the design process may start with the concept and then pass directly on to an analysis activity which helps to establish some geometric parameters that are then used in the subsequent development of a scheme. Alternatively, the design may start with the concept and proceed directly to the development of a scheme. In this case the analysis processes are employed at a later stage, in a checking role, to confirm that the proposed scheme does in fact meet the original requirements and specification.

Both of these methodologies are used extensively. The approach adopted is normally dependent upon the type of design being undertaken. In general terms over-constrained problems use the concept-analysis-scheming route whilst those that are under-constrained proceed in the order concept-scheming-analysis.

The analysis applications programs thus fulfil two main functions: the generation of basic design data and the confirmation of a proposed design scheme. Many of the applications programs on offer can be used in either mode but lend themselves more readily to use in a checking mode. The calculations and relationships for the correct design of a set of gear teeth, say, would be the same irrespective of whether they were used to allow a correct type to be selected or whether they were used to analyse a particular set that was presented. They are easier to incorporate into a program in the latter mode as the designer provides all the geometric parameters of the design and the program is left only with the calculations. Even the decision as to the acceptability of the design (based upon those calculations) is left to the designer. Programs designed to aid and advise the user in the correct selection of parameters (or even approaches) are very much more difficult to develop but are now starting to appear in greater numbers as expert systems techniques begin to be applied (see Module 8.3).

Most of the analysis programs that depend upon large and complex computational activities thus tend to be used primarily in the verification role, using existing information in the geometric database; typical examples are finite element and electronic circuit analysis. There are however also programs that are used essentially to generate graphics; typical of these are the programs used for the kinematic modelling of machine parts and the laying out of electrical printed circuit board designs.

FINITE ELEMENT ANALYSIS

The finite element method is normally associated with the analysis of large and complex structures, especially in the aircraft industry. Here the technique allows the load-carrying structure of the complete aircraft to be analysed under various loading conditions that can range from the static loads incurred whilst taxiing, through to the vibrations and shock loadings that could occur during extreme conditions in flight. The technique can, however, be applied to the solution of a diverse range of other engineering problems. Successful areas of application include fluid flow, lubrication, dynamic analysis, electromagnetic fields and heat flow. Tidal flows in an estuary can be modelled to find out how pollutants are dispersed or deposited upon the shore line. The distribution of temperature within a nuclear reactor can also be ascertained by the generation of an appropriate finite element model.

Although these examples may appear to deal with systems that are physically very different from each other, the solutions in all cases depend upon the ability to describe and resolve a complex set of differential equations. The finite element method addresses all these problems in the same general way, by using a discrete system to approximate a continuous one. The complex set of differential equations that model the physical situation are replaced by an approximate (but representative) set of algebraic linear equations which are solvable. This approach allows the continuous properties of the problem to be subdivided into 'patches' within which simplified properties and relationships are assumed. The equilibrium of the total system of patches is maintained at discrete points on the boundaries between them.

This process can be visualized if we imagine using a chain in which each link is a straight line in order to approximate the way in which a flexible cable would sag if hung between two posts. The links are the 'elements'. If the mass of each is known, as is the way in which it reacts with its neighbours, then the problem can be solved. The solution cannot accurately represent all positions on the cable. However the positions of the points (called the 'nodes') where the links join provide a satisfactory approximation to the sag of the cable.

This analogy also allows us to consider some of the extreme effects that the choice of element size can have upon the solution. If there are too few links in the chain then a very crude representation of the cable is provided and poor results will be obtained. If we subdivide these large elements into a great number of smaller ones then the approximate representation is closer to the true situation; but as each element can provide only an approximate representation (in this case a straight line) to the curvature of the cable at any point, then as the number of elements increases so do the errors incurred by the use of this simplification and, again, the results may be of little value.

Although finite element analysis programs can be arranged to provide information on the state and accuracy of the calculations as they proceed, it is usually left to the operator to determine how the model should be subdivided and to select the type and form of the elements. Thus, without an operator who has the requisite skill and knowledge, the solutions obtained may cost a lot, but be worth nothing.

In order to achieve successful results from a finite element program it is necessary to select the appropriate types of elements, arrange them into an appropriate mesh and, finally, set up the appropriate loading configuration. A finite element package associated with a CAD system may provide help, guidance or automatic selection for these activities. Basically the CAD system can provide, from its database, a full geometric description of the model to be analysed (assuming that it is a three-dimensional system). It is upon this that the mesh is generated (figure 8.1).

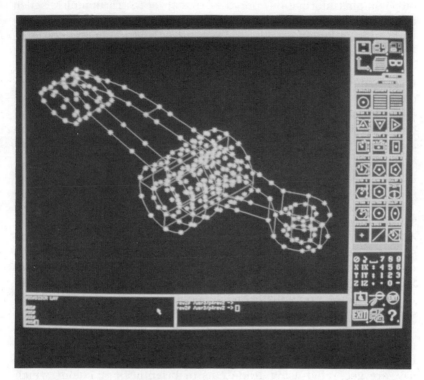

Figure 8.1 Finite element mesh.

Most systems provide automatic mesh generation. This does not, however, normally involve anything more than the automatic subdivision of a large space into a regular or irregular mesh of smaller elements, thus removing the need for the operator to specify every single element within a complex body. The system may provide a range of element types - extending from simple ones that represent beams, through those formed by triangles that can carry perhaps only constant strain conditions, to complex blocks with many nodes that can describe rapidly changing strain states. It is therefore necessary for the operator to understand where the greatest concentration of stresses is likely to occur within his design in order that he may choose the position and appropriate density of elements within that region. An optimum layout has few elements in areas of low stress and larger numbers in regions where high stress is expected.

The need for the user to make some prior estimate as to how the stresses are distributed, in order that a correct mesh can be selected, may cause difficulties for the inexperienced. Current finite element programs on CAD systems can provide very little help as the database only contains information upon the original geometric model. In order to provide any worthwhile guidance the system must also be provided with information about material properties and loading conditions. Once the mesh has been created interactively by the operator and any auto-meshing has taken place, the nodal description of the model must be down-loaded to a separate analysis package. This may be one of the many 'standard' finite element programs that can be bought commercially. Once the analysis has been done, the results can be transferred back to the CAD system and displayed. In some cases the actual analysis may be done on an entirely separate computer. Other systems have somewhat greater integration, with the analysis activity being run in a 'background mode' on the same computer whilst the graphical activities continue.

Once the analysis has been performed the data can be reviewed. This is most easily done with the aid of a graphical representation rather than by scanning sheets and sheets of numerical output. CAD systems offer a number of differing display techniques. When considering the deflection of a structure it is possible to redraw the model with the deflections of the nodes displayed at a greatly increased scale. Areas of greatest deflection are thus seen to bulge and sag when compared to the unloaded structure.

Other CAD systems also allow displays to be generated in which regions with similar states (for example stress or strain) are identified by a similar patterning or colouring. Contour maps of stress gradient, say, are generated using gradations of brightness or colour which allow regions of high intensity to be easily identified (figure 8.2).

The automatic generation and checking of meshes allows models to be analysed and evaluated graphically in a fraction of the time it would take to prepare and interpret the same data by hand. It is therefore possible to consider and evaluate more alternative designs, to a higher level of confidence, in the early stages of the design activity.

ELECTRONIC CIRCUIT ANALYSIS

The electronics industry has increasingly felt a need to be able to undertake detailed analysis of circuit designs before they are built. Historically, the industry was one in which 'for the cost of a handful of standard components', one could quickly throw together a representative prototype. This situation has, however, changed considerably in recent years. Advanced applications are now dominated by the use of digital systems that may be custom-made in the form of chips. The production of such devices is costly and time consuming and this means that components may not be available early enough for the construction of a prototype. There is, therefore, a need to simulate and

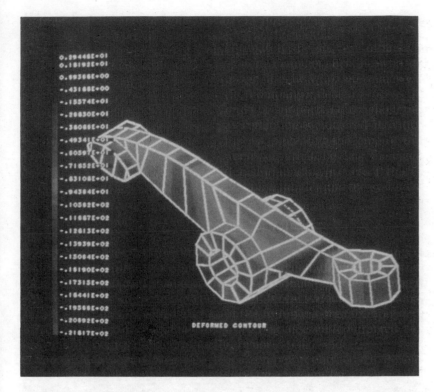

Figure 8.2 Display of finite element analysis results.

analyse the proposed design in order that errors can be corrected before the complex process of manufacture is undertaken.

The analysis of a circuit consisting of a few components with a limited number of interconnections can be easily undertaken manually. But once the number of elements starts to increase the problems rapidly multiply. With integrated circuits now being constructed that contain many thousands of separate devices, with perhaps millions of interconnections, the problem has grown way beyond the scope manual techniques. Moreover, the small size and compactness of these circuits creates other problems that need to be analysed. These include thermal effects within the devices, pick up of signals from adjacent devices or tracks and transient effects arising as a result of impulses created when the digital signals are switched on and off.

As in the case of finite elements, the range of integration and the sophistication of the programs on offer varies widely. At the most basic level there are simple, stand-alone packages that can analyse circuits constructed from standard resistive, capacitive, inductive and transistor elements. Such a program relies upon the CAD database for the description of the original circuit diagram, but all analysis activities are conducted within a separate suite of programs. The CAD system is simply used to create a file containing details of the devices and their connectivity (called a 'net list') that can be passed on to the analysis procedures. Figure 8.3 shows output from a typical CAD session.

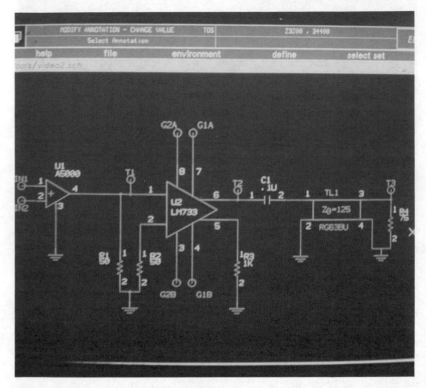

Figure 8.3   Electronic circuit design.

Two basic approaches have been employed in order to generate these kind of files. The circuit may be drawn using the standard drafting procedures, with the devices then being named to create a 'parts list' file. All connection lines are then tagged with a unique name and the names and pin numbers to which they connect. Depending upon the degree of integration achieved by the programs, some of this connectivity work may be performed automatically, leaving the operator to add in the links that are missing or deal with problems that cannot be resolved automatically.

The alternative approach is to use a circuit generation package that runs within the graphics environment. This allows the operator to select components (often from menus) and place them on a circuit layout. Each device is named upon insertion and all the connection pins are displayed. The circuit is then created by the operator adding 'connection lines' between the appropriate pins. The system, having laid down and tagged the connecting pins, is then able to provide the connectivity list automatically by identifying the pin numbers associated with each connecting line.

Such circuit generation programs often make provision for a low level of circuit verification to be carried out before the costly full analysis is undertaken. This checks the completeness of the circuit description and whether appropriate connections have been made. In this way, short circuits and unpowered or inactive devices can be identified and normal modes of operation can be verified. Major errors can thus be avoided, but errors created by subtle or inappropriate use of devices or interconnections still need to be detected by the skilled operator.

As the demands for this type of circuit analysis have increased so have the sophistication of the programs and the degree of integration with the CAD system. A whole new breed of CAD workstations has now appeared dedicated solely to this work. A range of special-purpose software can be run that allows circuits and devices to be assembled and simulated interactively on the system. Predefined dynamic representations of standard devices and circuits are available, as well as advice on device selection and fault finding procedures based upon expert systems.

With such systems the operator is able to specify the circuit and components and to indicate the power and state of the tests to be conducted. The system then performs the analysis, providing output information in a graphical form. Depending upon the degree of integration between the CAD system and the applications packages and the type of analysis being requested, the system may be able to provide a synchronized set of graphs for all the specified test points on the circuit. These allow logic errors to be identified and timing problems and transients ('spikes') to be observed. The use of such analysis programs can be a big help in eliminating many of the major problems that can lead to costly delays and reworking during the development of complex electronic circuits.

KINEMATIC MODELLING

In many design situations there is a need to investigate the changing relationships between various moving components. This may be necessary either when analysing a developed design or when selecting an appropriate design scheme that can then be detailed. During the design of, say, a windscreen wiping mechanism the designer must establish that any proposed solution successfully moves the pair of blades across the screen, that they do not interfere with each other and that the driving components do not clash with stationary parts of the body structure. Some parameters are thus established and a scheme developed. This can then be analysed to find out whether it meets all the other requirements. If it does not, then modifications are proposed (such as the changing of link lengths or pivot centres) and the analysis is repeated. Such a process continues, interactively, until a successful solution is reached.

Kinematic modelling is used to simulate the performance of a single body or mechanism throughout its working cycle. Such modelling techniques analyse the motion at discrete intervals of time and thereby provide an understanding of the system's performance that may include the velocities and accelerations occurring at the connections between component parts.

The assembly of those parts is defined by the restrictions that are placed upon the free motion of the bodies. If a body is unconstrained then it can be said to have six degrees of freedom (that is it can translate in any of the three orthogonal directions in space and rotate freely about these three axes). The kinematic model is assembled by stating how any two related elements are allowed to move with respect to one another. A 'rotary pair', for example, is one where movement is restricted to rotary motion in one plane, as when an arm swivels about a point on a body. In a similar way, 'pairs' can be defined to represent the linear extension of members, the sliding of a body upon a flat plane, the following of a cam profile, the helical motion of a nut down a screw thread, and so on. These pairs, together with the geometric positions of the various parts, allow the complete mechanism to be defined.

When performing such activities upon a CAD/CAM system the basic requirement is to move the parts, individually or in groups, with respect to one another to provide a simulation of their operation. The simplest kinematic modellers do just that. They allow the drawn components to be assembled into the correct mechanism arrangement which is then advanced or 'driven' into a series of new positions to represent the motion throughout the operating cycle. These representations are stored as separate 'frames' that can then be played back in sequence to give an illusion of motion. Such systems can be extended to provide numerical data on the changing positions of parts as the motion takes place.

These programs do not, however, handle the true kinematic integrity of the mechanism. Changes in position, size or attitude of one part could cause the linkage to dismember. Without the full kinematic representation or user intervention, the stored solution frames may be modified thus destroying the correct assembly. Full kinematics modellers start, not with the assembly of the drawn parts, but with the mathematical definition of the mechanism relationships. This is done by describing how one part attaches or relates to another part through the definition of a 'kinematic pair'.

There are two principal methods whereby the assembly can be analysed to obtain the positions of the parts throughout the mechanism cycle. The one most commonly used depends upon the relationships of all linkage or motion points being described by a series of trigonometric equations. The inverse transform (solution) of this set of equations then provides a description of the positions of the moving points that satisfies the equations at any chosen position. Programs based upon such techniques have been developed to handle both the kinematic and dynamic relationships of extremely complex mechanism problems. Figure 8.4 shows a mechanism assembled in this way.

Whilst these techniques allow mechanisms to be described by classical trigonometric equations, the inverse transform process does require the solution to be uniquely definable and it is also necessary that the accuracy required throughout the mechanism remains the same. In many cases this may not be so; in robotics, in particular (see Module 8.2), there may be many possible configurations that would allow a robot to reach a given position. In the case of a robot it is also obvious that a small error in rotation of the end effector can be ignored whilst the same angular error in the rotation of the base cannot. Another difficulty in employing the inverse transform technique on a commercial CAD/CAM system is that it is computationally demanding. It has to be redefined if changes in the geometry are made and the calculations must be repeated every time a new configuration is to be investigated.

The alternative approach is to achieve the solution by an iterative process. Basically, the technique is one in which the equations of kinematic integrity are expressed so as to describe the errors in assembling the kinematic pairs. The assumption is that when no errors are detected in the assembly of the pairs, then the total mechanism assembly is itself correct. If an error is detected when, say, a link is moved, then an iterative procedure is initiated that causes the errors to be corrected until the mechanism is again assembled in its proper form. These iterative processes use the known current state of the assembly to predict the new state by optimization techniques. The prediction is then checked for errors and a better solution attempted until the assembly is found to be satisfactory.

As these iterative processes seek a solution that allows the kinematic integrity to be maintained they are not dependent upon there being a unique solution to the problem. They simply find the 'first' solution and not all possible solutions. They thus have the advantage that different answers can be produced by moving the

Figure 8.4 Assembly of a mechanism.

assembly to different initial configurations. As they depend upon a process of prediction and correction, they can be built to handle very complex problems in a truly creative manner. The difficulties in defining these relationships and improving the iteration procedures are currently limiting the application of such techniques. The development of knowledge-based and constraint-modelling techniques (see Module 8.3) will allow them to be more widely applied in the future.

PRINTED CIRCUIT BOARD LAYOUT

Once an electrical circuit consisting of discrete components has been designed (possibly with the aid of a circuit analysis package described earlier), there is a need to decide how best to lay it on a circuit board in order to ensure that manufacture is as easy as possible. The best arrangement is normally one in which all the connecting tracks of the board pass directly from one component connecting point to another without having to cross any other tracks. As the number of components placed upon a single board increases the chances of finding a layout that provides such simple connectivity by manual methods is very remote (indeed be no such layout may be possible). Many different arrangements of components may have to be tried before a satisfactory one is found.

Moreover, in order to ease the existing problems (and to allow an even greater density of the new surface mounted components to be accommodated), circuit board techniques have advanced far beyond the single layer of tracks used in the early days. Many boards are now produced with multi-layers of tracks using both sides of the boards, with connections between the sides being provided by through-plated holes. The connectivity path between two points may thus cross over other tracks, by passing to a different level, many times along its route.

Such techniques, whilst allowing more components to be placed upon the board, present the layout designer with a mind-boggling three-dimensional puzzle.

Special purpose programs have been developed to aid the designer in the complex task of optimizing the layout configuration. All systems start with the connectivity net list as used in the circuit analysis package. Depending upon previous design activities, this may or may not already exist within the system in graphical form. The usual approach is to start with the designer's suggested layout. The connectivity lines are reduced to single straight lines joining the required set of pin connectors. This results in a spiderwork of crossing lines which must, ideally, be untangled to eliminate all cross-overs.

The algorithms employed by the system vendors are closely guarded secrets. Each vendor hopes to improve upon the methods used by their competitors and so gain an edge in the marketplace. Whilst the degree of sophistication varies considerably between packages, they are all based upon the idea of looking at the limited number of options that exist when two crossing wires must avoid each other. Either one line must be rerouted around an end of the other line or, if it is possible for the lines to avoid each other by 'jumping' to a different layer, then that possibility must also be considered.

For a given arrangement of components there is a limited number of possible routes for each crossing pair on the board. However, due to the large number of possible crossing pairs and their interdependency, a very large search may be needed in order to find the optimum circuit routing with the minimum of unavoidable bridging wires. Various strategies can be adopted to aid the solution, including repositioning of the components and starting with the one that has the minimum of crossing pairs. Some systems progressively search through all combinations (given time), whilst simpler ones rely upon the operator providing the various configurations that are to be investigated. A successful layout is shown in figure 8.5.

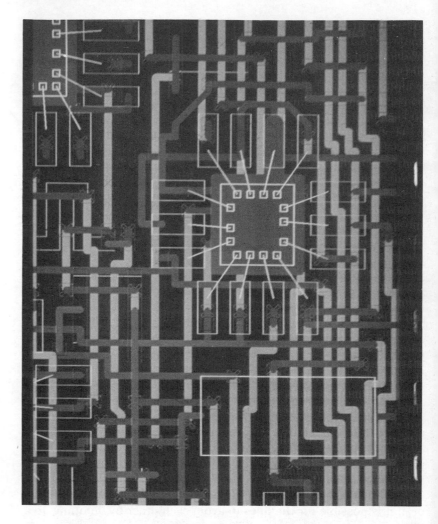

Figure 8.5 Printed circuit board layout.

As an attempt to search all possible configurations to find an optimal solution can be a long and costly operation, most systems do not guarantee to undertake an exhaustive search. Instead they attempt to 'unravel' the majority of the cross-overs and leave a small residue for the designer to sort out. This usually means investigating winding a track by a circuitous route around many of the components. The techniques are advancing and more sophisticated strategies are being applied (often built around an expert system). As a consequence, the speed of operation is being improved and the size of unresolved areas of the layout is being reduced.

MODULE 8.2 MANUFACTURING-CENTRED APPLICATIONS PROGRAMS

The nature of the design process is such that the details of the design are formalized during the iterative process linking the concept, scheming and analysis activities (see figure 1.8). Once the design is fully developed the downstream activity of manufacturing is often regarded as simply one of turning the completed design into a reality. This approach assumes that the design is correct and appropriate for manufacture and that production planners neither have any opinions on the design nor wish to make any comments (a situation rarely met in practice!).

This concept of unidirectional communication, with data flowing only *from* design *to* manufacture, is often reflected in the form of the current applications programs. These assume that once design is complete, it is passed to the downstream processes where it can be manipulated into a form appropriate to enable production. It is further assumed that all comments relating to manufacturing requirements have already been taken into account, or that no such comments need be considered. Such applications programs are therefore primarily geometric model manipulators (to which additional 'properties' may be added interactively by the operator).

Once a part has been fully developed within the geometric database many downstream processes can be planned on a semi-automatic basis. If the part is to be machined from a solid piece of material, the movements the cutter must go through in order to perform the various profiling and boring operations can be calculated as a dependent geometric operation. The final track of the toolpath can also be fully described within the original geometric database. Similarly the cavity of a moulding tool or casting pattern can be considered to be basically the inverse geometric form of the original model held in the database. The blank shape of a folded metal part can be thought of as the correct assembly of different elements or views drawn from the original part.

Another class of geometric activities is involved when a collection of models are to be nested together within a defined geometric space in order to establish the best arrangement that produces the minimum of scrap material. This is a task that may be necessary, for example, when planning the arrangement of parts to be pressed or routed from metal sheets or when cutting patterns from sheets of fabric or plastic.

It is in aiding these sorts of activities that a CAD/CAM system starts to produce real benefits. Only rarely is it possible to show that using a system simply in the production of engineering drawings will generate significant savings - all too often the time taken to produce a drawing with a CAD system is greater than if had been constructed manually. It is when the data held in the database can be used in generating and controlling downstream manufacturing processes that the major advantages of CAD become evident.

There are many operations which manipulate the original geometric database to perform downstream activities, and they are highly system- and process-specific. It is therefore impossible, within the scope of a book such as this, to provide anything other than a few broad examples to illustrate how such applications packages typically work. Those discussed have been chosen to reflect operations found in both electrical and mechanical engineering (the examples being CAM and robotic assembly) and to demonstrate the level of process analysis that can be carried out upon a CAD workstation (exemplified by applications packages for mould design).

COMPUTER-AIDED MANUFACTURE

When the drafting/detailing stages of a design have been carried out using a CAD system the user has built a computer model of the component concerned. Care should be taken to ensure that as well as producing a satisfactory engineering drawing, the data has also been created and stored in a form which allows it to be used in aiding subsequent activities.

The most common example of this is the use of the CAD model in computer-aided manufacture (CAM). Put at its simplest, the idea of CAM is to translate the computer CAD model of the component into a physical model. However, it is not that easy to convert the boundaries of the computer model directly into numerically controlled (NC) instructions for a machine tool. The restrictions of the model definition and the capabilities of the machine tool and its cutter need to be taken into account. In order to do this the CAM package must have additional information and must be capable of interaction with the user.

Consider a specific example, the process of 'pocketing', that is, forming a recess by removing material to a certain depth from a region specified by a plane boundary curve (figure 8.6). The CAD system may hold only information about the boundary itself and depth of the vertical surface to be created, it may not have a complete surface model of the hole. But it is not sufficient for the machine tool simply to follow the boundary, cutting to the specified depth, as this can leave material in the middle of the hole. (Conversely, if the hole is smaller than the cutting tool itself then the operation is impossible and should not be attempted.)

There are well-established algorithms for deciding how to remove material from a pocket once the boundary is known. So a simple way to provide this facility is to let the user give a pocketing command to the CAM system and then to indicate the bounding curve on the graphics screen. The system can then go ahead and produce the toolpaths to complete the operation. If the hole is very deep it will be necessary to proceed in steps and the user will have to indicate how much material it is safe to remove at each stage; this will depend on the feedrate and spindle rate of the cutter being used.

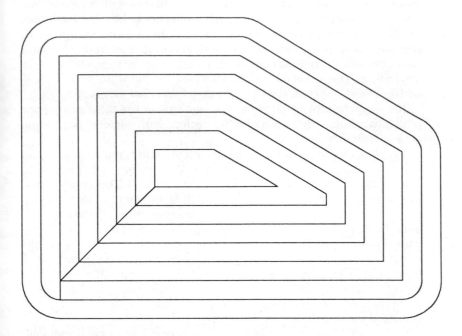

Figure 8.6   Pocketing.

A CAM package is thus a piece of software which allows the user to obtain NC machining instructions to produce the component. This software may be an integral part of the CAD software, it may be a distinct package on the same computer or it may even be resident on an entirely different machine. The less integrated the CAD and CAM systems are, the more problems are likely to arise in transferring information from one to the other. All systems need to identify and extract the relevant model geometry before the machining processes can be undertaken.

The main advantage of a CAM applications program lies in its ability to hold a repertoire of standard machining operations. The user of the system invokes a suitable one, indicates which part it is to be applied to and gives any extra information required by the system. This methodology combines the powers of the computer in extracting and manipulating geometric data from the CAD model and the expertise of the user in the machining technology. It also allows for the fact that the CAD model itself may not be complete; for example, the pocketing operation described above could be specified from a wire-frame model of the component.

If the CAD model is more sophisticated, then the CAM package can do more for itself. However there is also greater scope for things to go wrong and the need for user interaction is still present. Suppose that free-form surfaces defined in the CAD model are to be machined. This requires that the machine tool has movement in three axes at least, and preferably that it can also rotate the cutting head so that the tool cuts normally to the surface at all stages; this is termed 'five axis machining'. The common procedure is for the CAM package to deal with one surface patch at a time. While it can process this fairly easily (effectively the tool needs to sweep contours across the patch), there is always a danger that the tool will interfere with adjacent patches or other parts of the model.

Avoidance of such interference problems still remains mainly the responsibility of the user. Often a CAM system can provide a simulation of the machining operations (figure 8.7). This consists of a simple model of the machine tool head. It is shown in a large number positions, sequentially simulating the various cutting operations that must be undertaken, and so producing an animated cartoon of the complete process. The user can check visually for interferences and other problems. The user can also model the various clamps and fixtures in order to check that these will not be fouled. If a problem is seen to occur, then that part of the machining process needs to be redefined and a good CAM package must allow the user to change the details of the operations interactively.

Once a satisfactory simulation of the process has been produced, the CAM system can generate the precise instructions necessary for the machine tool to be used. As the instruction sets for different types of machine tool controllers vary considerably, it is usual for the CAM package itself to work in terms of a 'neutral' set of instructions. These are subsequently translated to the machine-specific form by a 'post-processor'. This is a piece of software resident within the CAM system and one is needed for each machine tool that is available. (Some systems have a 'generalized' post-processor which can be set up for any machine tool by providing detailed information about it.) The post-processed instructions are then sent to the tool. In the early days this was done by means of punched paper tape; but today the transfer is increasingly likely to be made by electronic means, with the CAM system and the machine tool being networked together.

The CAM information can also be used to estimate the time required to machine the component. This is calculated from the cutter paths that have been established and must take account of factors such as

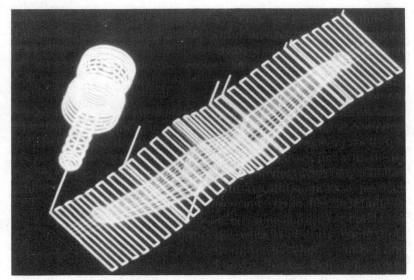

Figure 8.7 Simulation of machining operation.

the feed rates to be used and the number of times the cutter needs to be changed. This data can be fed into a process planning system which can then schedule the work of the various machine tools in a shop so that these are used to best advantage. This can mean that any given component can be produced on any one of a number of different machine tools, although this may involve the machining instructions having to be post-processed for each machine separately.

To summarize, a CAM package takes geometric information from the CAD model and combines this with expertise from the user to produce machining information. Again, work is being carried out into the use of expert and similar systems (see Module 8.3) in order to design systems that will rely less heavily upon the user. However the time when the user can just sit back and let the system do everything for itself seems still to be long way off.

ROBOTIC ASSEMBLY

The use of robots in manufacturing is becoming widespread. Initially, their main role was the performance of laborious tasks in hazardous environments. They were put to work on jobs such as loading large metal sheets into press tools, where noise levels and temperatures can be high and an operator error can easily result in maiming or death. In this type of operation the robots were only required to provide strength and reliability, but as their dexterity improved they increasingly found their way into assembly operations.

The most widely publicized robot application has been assembly in the motor industry. Here they have been employed to perform most operations, from welding, painting and assembly of windscreens through to visual inspection. As a result it is often assumed that it is in this type of industry that most robots are employed. In fact, they are being used for a wide, and rapidly increasing, range of other purposes. In laboratories they perform tasks such as taking samples from a large rack of specimens and systematically carrying out a range of tests and experiments. They are used to pick components and place them accurately upon a printed circuit board for assembly. They are employed, in conjunction with vision systems, to remove faulty components or damaged products from conveyor lines.

Robots are also becoming important elements within work cell environments. Here a number of items of machining or assembly equipment are grouped together to form a cell in which a particular set of activities is conducted or particular types of components are made. Each machine in the cell is programmed to perform a particular set of operations, but if the cell is to be completely autonomous the components must also be moved from one machine to another. Robots are the most flexible tools for this purpose, as, like the machine tools, they can switch from one program to another when the cell is required to produce a new type of component.

The effectiveness of the robot within its workplace depends directly upon the way in which it has been programmed. A program that brings it into collision with other parts of the cell is a disaster. One that involves slow, incorrect or wasteful motions reduces the efficiency, not just of the cell itself, but, perhaps, of the whole production line. It is therefore important that these machines are programmed in the most effective manner.

Traditionally, a robot has been programmed by means of a 'teaching' process in which an operator takes it, step by step, through the desired sequence of operations. This 'taught' activity is recorded and replayed whenever the robot is required to carry out the task. This approach requires either that the activity is performed within the actual robot workspace or that a replica is built. The latter option is costly, time-consuming and prone to errors whilst the former requires that production be halted for robot 'training' whenever a change is made.

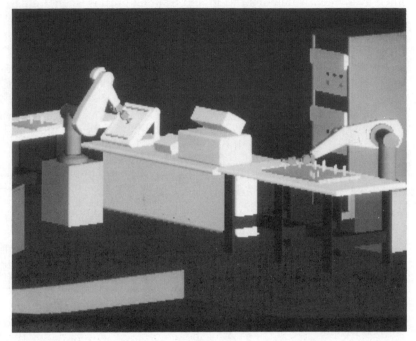

Figure 8.8 Manufacturing cell.

It is for these reasons the a number of applications packages are now being offered for use on CAD/CAM systems that can simulate robotic activities within a chosen working environment. These programs both allow the activities occurring in the work cell to be visualized and also enable the workings of the complete cell to be optimized.

The first function simply affords the user a way of checking, or demonstrating, how a cell is to perform. The cell designer can visually check that clashes do not occur and that the desired operations are performed correctly and in the right order. Such a visualization can then be used as an instructional aid for the team that is to lay out the cell and for those who are to supervise it. The simulation not only provides a high level of confidence that the process is going to work, it allows everyone to be aware of the critical areas that need to be watched and provides foreknowledge of mishaps that may be expected.

Such requirements can be provided by the simplest of kinematic modelling techniques (see Module 8.1). The creation of a series of 'frames' throughout the robot cycle allows the action to be replayed and checked for clashing with other elements in the cell and other incorrect operations. An animation of the required cell operation is then available for demonstration and training purposes (figure 8.8). If the robot simulation is to be used to optimize the work cell layout a higher level of modelling technique is required. It is then necessary to model not only the kinematics of the robot but also its response to, and dynamic interaction with, other mechanisms within the cell. This creates a major problem as the response and cell timing are ultimately dependent upon the form of the control algorithms used by the robot manufacturers. These algorithms are currently kept very secret as each manufacturer sees them as a means of maintaining a competitive edge over rival producers. In the absence of the full details, guesses have to be made which limit the accuracy of the simulation.

The use of a full robot simulation for the purpose of improving work cell efficiency has, however, proved to be very effective. It can be used to reduce cell cycle times, to check on the correct and effective selection of robots, and to improve the handling of the component parts and the layout of equipment within the cell. Some commercial packages are large and expensive programs to buy and operate, but even small savings on cell time within a large mass-production environment can result in substantial cost reductions on an annual basis.

MOULD DESIGN

When designing a component that is to be produced as a moulded plastic part, it is necessary to be aware of the problems associated with such a process. There are rules and good practice that can be used in an attempt to reduce the number of problems but, in the main, good designs depend upon the designer having an understanding of how the mould cavity is actually going to fill. Problems arise if there are voids or areas of only partial fill.

Often it is also necessary to attempt to lessen the amount of material used in order to reduce costs and to eliminate sinkages that are visible on the surfaces. It may also be a requirement that, for economy of manufacture, the parts should be produced in multi-cavity moulds. For small parts this may involve moulding up to 64 parts in one shot. The flow down the connecting tracks (called the 'sprue') must be arranged to allow all the cavities to be filled and yet be as economic as possible on the size of the sprue (some of which may be reclaimable). The balance between cavities must be correct and control of the material entry into the mould must be regulated if successful products are to result.

A limited number of special-purpose programs have been produced aimed at meeting the needs of mould designers. They are normally

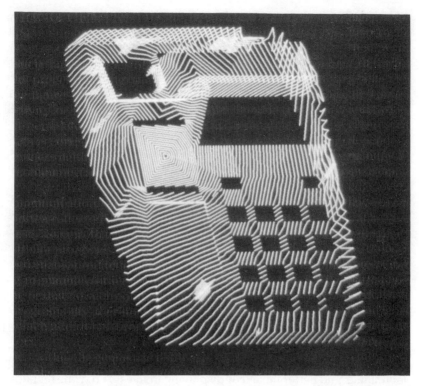

Figure 8.9   Simulation of flow within a mould.

separate analysis programs that have been coupled to a CAD system to take advantage of existing geometric models and of the system's display capabilities for presentation of the results.

Before any analysis can be performed it is necessary to extract the appropriate geometry from the system's database. After appropriate manipulation, this describes the full surface form of the cavity to be produced. The route and size of the connecting sprue and its entry point into the mould must also be described. The form of the entry point is usually chosen to be of a smaller diameter than the sprue, and possibly at an angle to the cavity, in order to create a higher velocity of injection into the mould and a reduced cross-section at which the component can be broken from the sprue. Once the components, their injection configurations and the sprue have all been designed and assembled then a flow analysis can be undertaken.

The analysis of the flow into the mould involves making assumptions as to how the chosen material flows under differing conditions of pressure and temperature. If these assumptions are accurate it is possible to determine how far the material will have flowed into the mould after a given time under any set of input conditions. As the flow proceeds, the temperature and pressure drop and so progressively slow down the motion of the material already in the mould. Once a critical set of conditions has been reached the material ceases to flow and leaves the remaining parts of the mould unfilled.

The graphics displayed by these analysis packages allow the designer to watch how the material flows through the spaces (figure 8.9). Voids, possibly due to using too little material, when attempting to fill the mould, can be observed. Changes in properties or the slowing of the material front can provide indications of problems that could occur when using the mould as designed. Changes can then be made to the sprue, the injection point or the cavities in an attempt to relieve these effects.

By analysing the mould performance, and redesigning where necessary, it is possible to improve greatly the performance of a chosen mould design. This may not only reduce the scrap rate but may also help the designer to economize on the amounts of material used. These techniques can greatly improve the productivity of the whole manufacturing process.

MODULE 8.3 CONTROL-CENTRED APPLICATIONS PROGRAMS

The initial emphasis in the development of CAD/CAM systems was towards the creation of ever more complex geometric modelling techniques and better means of displaying images. Today the technology has become centred upon the modelling database, with analysis activities interacting with it and manufacturing activities extracting the data they require from it.

The development of large, centralized systems and of others based upon networks of distributed graphics workstations has, for different reasons, led to a realization that the design process needs a high degree of control. A very large system handles and stores such quantities of data that the process of updating them, controlling their issue, and handling the authorization of hundreds of thousands of drawings is becoming an enormous task.

The control may in fact be provided by having a single copy of each design upon a single computer. However, the total system may have grown so large that a user may be forced to communicate with a remote machine in order to retrieve a desired part. Once it is located, he may be denied access to the part by that machine's security system. With a distributed system of workstations the degree of control provided may be less, and the data may be created and held locally by the user. This can result in a number of similar parts being held, at different levels of completeness, by different people on the same system. Without some form of management control over the design process chaos can quickly result.

Many attempts are now being made to include data management programs within CAD/CAM facilities. These are directed towards establishing the relationships between drawings within assemblies and piece-parts. Records are kept of developments, changes and of issue status and authority. Whilst these approaches go a long way towards achieving management control within the design department, in a similar manner to that previously achieved by the old drawing office log, they provide little or no control over the many downstream operations that depend upon this data. The status of the data must be known and the authority to proceed must be acquired before any of the operations can be carried out and the results applied within a manufacturing environment.

Some CAD vendors provide standard production planning and management programs that can be run on their systems. As these rarely integrate with the design database and only use the graphics facilities to display results in chart or flow diagram form, they are virtually identical to those covered in any text on production techniques and therefore are not considered in this book.

The major difficulty in providing the necessary controlling functions for CAD/CAM systems seems to be the attitude that they should be provided solely for the CAD/CAM system itself. If, however, one recognises that geometric modelling and manipulation techniques are but individual aspects of a total design process then the solution is obvious: CAD/CAM should be a function that is integrated within an overall design management system. It is no more than one sub-activity that must be carried out within a total design process.

This is an attitude that is gradually taking root, but the speed of its introduction depends upon developments currently taking place in expert or knowledge-based systems. Only with these new techniques can the complex and ill-defined process of design be handled and integrated in a creative manner.

EXPERT SYSTEMS

Expert system programs were developed to allow problems of logic, as opposed to the numerical manipulations that are the basis of conventional computer programs, to be addressed. The programs operate by searching through a large number of facts to establish whether a proposition is true or not. In order to be able to test a proposition, it is necessary to write the facts in the form of rules that can be applied and from which conclusions can be drawn. A collection of facts could be assembled, for instance, detailing various airline schedules and the system could then be asked whether it was possible to fly between various locations, arriving by specified times. The program would be required to search systematically through the flight details and transfer possibilities, until a combination that satisfied all the requirements was found or it was established that the specified conditions could not be met.

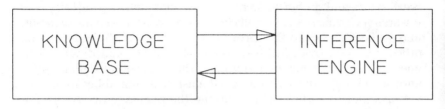

Figure 8.10   Simple expert system architecture.

The rules usually appear within computer programs in the form of IF-THEN-ELSE type of statements. In a simple system, facts and items of partial knowledge can be considered as nodes in a tree diagram. These are connected by edges which indicate how deductions can be drawn when other items have been decided. Among the nodes are the various conclusions that the system is able to draw. By finding out what is known (that is, by asking questions of the user), the system can work its way through the tree until a desired conclusion is established - this is 'forward chaining'. Alternatively, 'backward chaining' may be used, in which case the system starts from a conclusion whose truth is to be decided and works backwards through the tree to see what information is required to make the decision.

A feature of expert systems which is often regarded as important is that the knowledge and the rules for a particular application can be distinguished from the program that handles them. The program itself is called the 'inference engine' and can handle rules for any situation, provided they are supplied in a recognizable form. The rules and the known facts comprise the 'knowledge base' and are the data upon which the inference engine program operates (figure 8.10).

Such logical processes can be applied to assist in the solution of many engineering problems. Programs have been written that allow hydraulic components to be assembled; one has even been reported that can design gearboxes. Others are being developed to assist in the creation of machine cutter information for NC tools. Jig and tool designs can be selected by these techniques and the arrangement of components within an electronics assembly can be similarly chosen. All of these cases can be handled by an expert-system approach as long as the data can be reduced to a (preferably small) set of logical rules.

This works well in applications which can be fully described in terms of the rule structures available. In particular, expert systems are good at handling fault-finding and diagnosis-type tasks. The application must be very well understood and design rules must exist to cover all cases. In cases where nobody really knows precisely what the rules are, certain empirical 'black magic' guidelines have been used instead. The shapes used on preforming roller dies in the forging industry are an example of this; it does not matter greatly what shape is given to the preform provided it has the same volume at each stage. While expert systems can be used in these kinds of area, it is not clear that they offer any particular advantage over a special-purpose analysis program or, in a CAD context, parametric programs (see Module 6.4).

As was shown in Module 1.2, design is an intricate process of trying to reconcile conflicting constraints. It is here that expert systems should have the greatest potential. The constraints could themselves be regarded as rules. However, there is rarely a unique solution to a design problem. A real component is described by real numbers. The current expert system shells tend to deal with either 0 and 1 (that is true and false) or just with the integers. The tree structure holding the rules can deal with only a finite number of possibilities, while even a simple rule such as a particular dimension lying within a given tolerance band, has an infinite number of allowable solutions.

The ideas of forward and backward chaining through the rules do not work well in the context of general design. Suppose we wish to design a telephone handset and that all the rules for so doing are known. Forward chaining would presumably try to generate all the infinite variations on the handset shape. Backward chaining is equally invalid; it assumes you know what the answer is before you start and if this were the case there would be no design problem at all!

What seems to be required is a different version of an expert system in which the 'rules' are the 'constraints' upon the design. It is much easier to say what is wrong with an inadequate design than to generate an acceptable one. The constraints describe what cannot be done. The inference engine would then look at the given constraints and try to select the best compromise design solution. This avoids the generation of an infinite number of possibilities.

However the search process used would need to involve some kind of optimization technique using real numbers, rather than graph-searching algorithms related to integer values. The solution such a system proposed might not be correct. If it was not, then the user ought to be in a position to say why it was not acceptable and this information should form a new constraint to be added to the system before a new search was made. Thus the system and the user would learn about the problem together. The rule base would be built up as the best design was found. It could then be stored and re-used when a similar problem arose. In this way knowledge about particular application areas would be generated and stored. This type of approach would also seem to use the differing talents of man and machine to best effect.

Currently available rule-based expert systems are seen to be capable of handling certain very specific areas. They attempt to encapsulate the knowledge of one or more experts so that it can be accessed by others. However the diversity of application areas in design is so large that it is necessary to develop advanced systems, based upon the constraint-modelling approaches, before many of the more difficult problems can be addressed. It is systems based on these techniques, working hand in hand with the creative human designer, that seem likely to offer the best way forward towards the truly integrated and intelligent CAD/CAM systems of the future.

FURTHER READING

The following list of publications is not intended to represent a complete bibliography of CAD and related areas. It aims to provide suggestions for background and further reading. The items have been arranged into groups under the sections of this book to which they relate (some items appear in more than one group). The first group are publications of general interest.

General

Aleksander, I and Burnett, P, *Reinventing Man*, Kogan Page, London, 1983

Andreasen, M M and Hein, L, *Integrated Product Development*, IFS/Springer Verlag, 1987

Besant, C B and Lui, C W K, *Computer-Aided Design And Manufacture*, Ellis Horwood, Chichester, 1986

Bezier, P, *Numerical Control, Mathematics and Applications*, Wiley, New York, 1972

Emery, F E, *Systems Thinking*, Penguin Books, London, 1969

Encarnacao, J L and Schlechtendazhl, E G, *Computer-Aided Design*, Springer Verlag, New York, 1983

Groover, M P and Zimmerman, E W, *Computer-Aided Design and Manufacturing*, Prentice-Hall, Englewood Cliffs, N.J., 1984

Hatavany, J, 'Computer aided design', *Computer-Aided Design*, 16 (1984) 161-165

Holmes, R, *The Characteristics of Mechanical Engineering Systems*, Pergamon, Oxford, 1977

Michie, D, *On Machine Intelligence*, Edinburgh University Press, 1974
Milner, D A, *Computer-Aided Engineering for Manufacture*, Kogan Page, London, 1986
O'Grady, P J *Controlling Automated Manufacturing Systems*, Kogan Page, London, 1986
Warner, M, (ed), *Microprocessors, Manpower and Society*, Gower Publications, London, 1984

Section 1 The Design Process

Alger, J R M and Hays, C V, *Creative Synthesis in Design*, Prentice-Hall, Englewood Cliffs, N.J., 1964
Campbell, D, *Take the Road to Creativity and get off Your Dead End*, Argus Communications, 1977
Hubka, V, *Principles of Engineering Design*, Butterworth, Guildford, 1982
Mayall, W H, *Principles in Design*, Design Council, London, 1979
Medland, A J, *The Computer-Based Design Process*, Kogan Page, London, 1986
Medland, A J, and Burnett, P, *CADCAM in Practice*, Kogan Page, London, 1986
Pitts, G, *Techniques in Engineering Design*, Butterworth, Guildford, 1973
Ray, M S, *Elements of Engineering Design*, Prentice-Hall, Englewood Cliffs, N.J., 1985
Shoup, T E, Fletcher, L S and Mochel, E V, *Introduction to Engineering Design*, Prentice-Hall, Englewood Cliffs, N.J., 1981

Section 2 System Configurations

The best references here are manufacturers' information sheets and manuals. Some information is available in more general texts, the following being examples.
Besant, C B and Lui, C W K, *Computer-Aided Design and Manufacture*, Ellis Horwood, Chichester, 1986
Groover, M P and Zimmerman, E W, *Computer-Aided Design and Manufacturing*, Prentice-Hall, Englewood Cliffs, N.J., 1984
Voisinet, D D, *Introduction to Computer-Aided Drafting*, McGraw-Hill, New York, 1985

Section 3 Entity Descriptions

Bohm, W, Farin, G, Kahmann, J, 'A survey of curve and surface methods in CAGD', *Computer-Aided Geometric Design*, 1 (1984) 1-60

Coons, S A, *Surfaces for Computer Aided Design of Space Forms*, Report MAC-TR-41, MIT, 1967

Cox, M, G, 'The numerical calculation of B-splines', *J.I.M.A.*, 10 (1972) 134-149

de Boor, C, 'On calculating B-splines', *J. Approx. Th.*, 6 (1972) 50-62

de Boor, C, 'Package for calculating with B-splines', *SIAM J. Num. Anal.*, 14 (1977) 441-472

de Casteljau, F, *Outillage Methods Calcul*, André Citroen Automobiles SA, Paris, 1959

de Casteljau, F, *Shape Mathematics and CAD*, Kogan Page, London, 1985

Faux, I D and Pratt, M J, *Computational Geometry for Design and Manufacture*, Ellis Horwood, Chichester, 1979

Ferguson, J C, 'Multivariate curve interpolation', *J.A.C.M.*, 11 (1964) 221-228

Foley, J D and Van Dam, A, *Fundamentals of Interactive Computer Graphics*, Addison-Wesley, Reading, Massachusetts, 1982

Forrest, A R, 'On Coons' and other methods for the representation of curved surfaces', *Comp. Graph. and Image Proc.*, 1 (1972) 341-359

Garden, Y and Lucas, M, *Interactive Graphics in CAD*, Kogan Page, London, 1985

Giloi, W K, *Interactive Computer Graphics*, Prentice-Hall, Englewood Cliffs, N.J., 1978

Gordon, W J and Riesenfeld, R F, 'B-spline curves and surfaces', in *Computer-Aided Geometric Design*, (Barnhill R E and Riesenfeld R F, eds), Academic Press, 1974

Mullineux, G, 'Approximating curve shapes using parameterized curves', *I.M.A. J. App. Math.*, 29 (1982) 203-220

Mullineux, G, 'Reducing the degree of high order parameterized curves', *Proc. CAD82 Conf.*, 1982

Mullineux, G, 'Surface fitting using boundary data', *Proc. CAD84 Conf.*, 1984

Newman, W and Sproull, R, *Principles of Interactive Computer Graphics*, McGraw-Hill, 1979

Section 4 View Transformations

Foley, J D and Van Dam, A, *Fundamentals of Interactive Computer Graphics*, Addison-Wesley, Reading, Massachusetts, 1982

Newman, W and Sproull, R, *Principles of Interactive Computer Graphics*, McGraw-Hill, 1979

Rawson, E G, 'Vibrating varifocal mirrors for 3D imaging', *IEEE Spectrum*, September 1969, 37-43

Section 5 Types of CAD Modelling Systems

Baer, A, Eastman, C, and Henrion, M, 'Geometric modelling: a survey', *Computer-Aided Design*, 11 (1979) 253-272

Sutherland, I E, Sproull, R F and Schumacker, R A, 'A characterization of ten hidden-surface algorithms', *Comp. Surv.*, 6 (1974) 1-55

Section 6 The User Interface

Gonzalez, J C, Williams, M H and Aitchinson, I E, 'Evaluation of the effectiveness of Prolog for a CAD application', *IEEE CG&A*, (1984) 67-74

Huckle, B, *The Man-Machine Interface*, Savant Institute, 1981

Jones, P F, 'Four principles of man-computer dialogue', *Computer-Aided Design*, 10 (1978) 197-202

Martin, J, *Design of Man-Computer Dialogues*, Prentice-Hall, Englewood Cliffs, N.J., 1973

Meadow, C T, *Man-Machine Communication*, Wiley, New York, 1970

Swinson, P S G, 'Logic programming: a computing tool for the architect of the future', *Computer-Aided Design*, 14 (1982) 97-104

Section 7 System Effectiveness and Organization

Encarnacao, J L, Torres, O F F and Worman, E A, *CAD/CAM as a Basis for the Development of Technology in Developing Nations*, North Holland, Oxford, 1981

Medland, A J and Burnett, P, *CADCAM in Practice*, Kogan Page, London, 1986

Section 8 Applications and Extensions

Aleksander, I, *Computing Techniques for Robots*, Kogan Page, London, 1985

Alty, J L and Coombs, M J, *Expert Systems, Concepts and Examples*, N.C.C. Publications, 1985

Aquesbi, A, Bocquet, J C, Fouct, J M, Tichkiewitch, S Reynier, M and Trau, P, 'An expert system, for computer aided mechanical design', *Proc. Information Processing 83*, Mason (ed), 1983

Basden, A, "On the application of expert systems", *Int. J. Man. Machine Studies*, 19 (1983) 461-477

Bathe, K-J, *Finite Element Procedures in Engineering Analysis*, Prentice-Hall, Englewood Cliffs, N.J., 1982

Bonnet, A, *Artificial Intelligence − Promise and Performance*, Prentice-Hall, Englewood Cliffs, N.J., 1985

Brown, D C and Chandrasekaran, B, 'Expert systems for a class of mechanical design activity', *Proc. Working Conf. on Knowledge Engineering in CAD*, Budapest, 1984

Cholvy, L and Foisseau, J, 'Rosalie: A CAD object-oriented and rule-based system', *Proc. Information Processing 83*, Mason (ed), 1983

Colyer, B and Trowbridge, C W, 'Finite element analysis using a single user computer', *Computer-Aided Design*, 17 (1985)142-148

Cowan, D F, 'CAD company caters for the electronics industry', *Computer-Aided Design*, 17 (1985) 266-270

Dorf, R C, *Robotics and Automated Assembly*, Reston Publishing Company, Virginia, 1983

Dym C L, 'Expert systems: new approaches to computer-aided engineering', *Engineering with Computers*, 1 (1985) 9-25

Fenner, R T, *Finite Element Methods for Engineers*, Macmillan, London, 1975

Gero, J S (editor), 'Special issue: expert systems', *Computer-Aided Design*, 17 (1985) 395-468

Johnson, L and Keravnou, E T, *Expert Systems Technology – A Guide*, Abacus Press, 1985

Kitajima, K and Yoshikawa, H, 'HIMADES-1: a hierarchical machine design system based on the structure model for a machine', *Computer Aided Design*, 16 (1984) 299-307

Knowles, N C, 'Finite element analysis', *Computer-Aided Design*, 16 (1984) 134-140

Ko, H and Lee, K, 'Automatic assembling procedure generation from mating conditions', *Computer-Aided Design*, 19 (1987) 3-10

Kochan, D (editor), *Integration of CAD/CAM*, Elsevier, Amsterdam, 1984

Lee, K and Andrews, G, 'Inference of the positions of components in an assembly: part 2', *Computer-Aided Design*, 17 (1985) 20-24

McLeod, N D, 'Computers are putting Australia on the map', *Computer-Aided Design*, 17 (1985) 271-272

Medland, A J and Mullineux, G, 'The investigation of a rule-based spatial assembly procedure', *Proc. CAD86 – Knowledge Engineering and Computer Modelling in CAD*, 1986

Mejtsky, G J and Rahnejut, H, 'Introducing GPSL: a flexible manufacturing simulator', *Computer-Aided Design*, 17 (1985) 219-224

Molian, S, *The Design of CAM Mechanisms and Linkages*, Constable, London, 1968

Mullineux, G, 'Optimization scheme for assembling components', *Computer-Aided Design*, 19 (1987) 35-40

Owen, T, *Assembly with Robots*, Kogan Page, London, 1985

Patten, W J, *Kinematics*, Reston Publishing Company, Virginia, 1979

Przemieniecki, J S, *Theory of Matrix Structural Analysis*, McGraw-Hill, New York, 1968

Redford, A and Lo, E, *Robots in Assembly*, Open University Press, Milton Keynes, 1986

Ross, C T F, *Finite Element Methods in Structural Mechanics*, Ellis Horwood, Chichester, 1985

Sell, P S, *Expert Systems – A Practical Introduction*, Macmillan, 1985

Swift, K G, Matthews, A, Firth, P, and Syan, C S, 'Artificial intelligence in engineering design', *Proc. Int. Conf. Man or Machine – A Choice of Intelligence*, 1984

Swinson, P S G, 'Prolog: a prelude to a new generation of CAAD', *Computer Aided Design*, 15 (1983) 335-343

Thomson, C C, 'Robot modelling – the tools needed for optimal design and utilization', *Computer-Aided Design*, 16 (1984) 335-337

Winston, P H, *Artificial Intelligence*, Addison-Wesley, 1984

Wolfendale, E, *Computer-Aided Design of Electronic Circuits*, Illife Books, London, 1968

Wordenweber, B, 'Finite element mesh generation, *Computer-Aided Design*, 19 (1984) 285-291

Zienkiewicz, O, *The Finite Element Method*, McGraw-Hill, Maidenhead, 1977

INDEX